AF357829

Recent Clinical and Basic Research on Endocrine Surgery

Recent Clinical and Basic Research on Endocrine Surgery

Editor

Jin Wook Yi

Basel • Beijing • Wuhan • Barcelona • Belgrade • Novi Sad • Cluj • Manchester

Editor
Jin Wook Yi
Inha University Hospital and
Inha University
Incheon
South Korea

Editorial Office
MDPI AG
Grosspeteranlage 5
4052 Basel, Switzerland

This is a reprint of articles from the Special Issue published online in the open access journal *Medicina* (ISSN 1648-9144) (available at: https://www.mdpi.com/journal/medicina/special_issues/ Q1UNIS02D7).

For citation purposes, cite each article independently as indicated on the article page online and as indicated below:

Lastname, A.A.; Lastname, B.B. Article Title. *Journal Name* **Year**, *Volume Number*, Page Range.

ISBN 978-3-7258-2627-8 (Hbk)
ISBN 978-3-7258-2628-5 (PDF)
doi.org/10.3390/books978-3-7258-2628-5

Contents

About the Editor

Jin Wook Yi

Associate Professor, Inha University Hospital and Inha University College of Medicine. Director of International Cooperation Division Chief of Robot Surgery Center

Professor Jin Wook Yi is a distinguished surgeon and academic based at Inha University Hospital, where he is a faculty member in the Department of Surgery. With a specialization in endocrine surgery, he has built a reputation for his expertise in minimally invasive and robotic surgery.

In addition to his clinical practice, Professor Yi is actively engaged in medical research, contributing to advancements in clinical oncology and biomedical informatics. His work has been published in various peer-reviewed journals, and he has given presentations at many national and international conferences. He has also held live surgery workshops for endoscopic and robotic thyroid surgery in many countries, including Georgia, India and the Philippines.

Professor Yi is also dedicated to medical education, mentoring both undergraduate and postgraduate students in endocrine surgery. His approach emphasizes not only technical proficiency but also the development of a compassionate, patient-centered approach to surgical care and active research in basic medical sciences.

Article

Comparative Analysis of Laparoscopic and Robotic Transperitoneal Adrenalectomy Performed at a Single Institution

Yun Suk Choi [†], Ji Sun Lee [†] and Jin Wook Yi [*]

Department of Surgery, Inha University Hospital, College of Medicine, Incheon 22332, Republic of Korea
* Correspondence: jinwook.yi@inha.ac.kr; Tel.: +82-32-890-3437; Fax: +82-32-890-3549
† These authors contributed equally to this work.

Abstract: *Background and Objectives:* Laparoscopic adrenalectomy (LA) is the standard surgical approach for adrenalectomy. At present, robotic adrenalectomy (RA) has been introduced at various hospitals. This study evaluated our initial experience with robotic adrenalectomy compared with conventional laparoscopic adrenalectomy. *Materials and Methods:* From October 2018 to March 2022, 56 adrenalectomies were performed by a single endocrine surgeon. Thirty-two patients underwent LA (LA group), and twenty-four patients underwent RA (RA group). *Results:* Patients in the RA group were significantly younger than those in the LA group (48.6 ± 9.7 years vs. 55.1 ± 11.4 years, $p = 0.013$). The RA group had a shorter operation time than the LA group (76.1 ± 28.2 min vs. 118.0 ± 54.3 min, $p < 0.001$). The length of hospital stay and postoperative pain level between the two groups were similar. There were no complications in the RA group. There was no significant difference in the pathologic diagnosis between the two groups. The cost of surgery was significantly higher in the RA group than in the LA group (5288.5 US dollars vs. 441.5 ± 136.8 US dollars, $p < 0.001$). *Conclusions:* In our initial experience, RA showed a shorter operation time than LA and no complications. RA could be a viable alternative surgical option for adrenalectomy, notwithstanding its higher cost.

Keywords: robotic adrenalectomy; minimally invasive surgery; adrenal gland neoplasm; laparoscopic adrenalectomy

Citation: Choi, Y.S.; Lee, J.S.; Yi, J.W. Comparative Analysis of Laparoscopic and Robotic Transperitoneal Adrenalectomy Performed at a Single Institution. *Medicina* **2022**, *58*, 1747. https://doi.org/10.3390/medicina58121747

Academic Editor: Žilvinas Dambrauskas

Received: 10 November 2022
Accepted: 27 November 2022
Published: 29 November 2022

Publisher's Note: MDPI stays neutral with regard to jurisdictional claims in published maps and institutional affiliations.

1. Introduction

The adrenal glands are mustard-colored paired organs located in the superomedial part of the kidney in the retroperitoneum. They secrete various kinds of steroid hormones, such as cortisol and aldosterone from the cortex and catecholamine from the medulla. An adrenalectomy is recommended for malignancies such as adrenocortical cancer, size-increasing adrenal tumors, and functional adrenal tumors such as those causing primary aldosteronism, Cushing's syndrome, and pheochromocytoma [1].

Adrenal surgery has a long history with steady progress. The first adrenal surgery was performed by Thorton in 1889, wherein an adrenal mass was described as a large sarcoma with a left suprarenal capsule [2]. In 1914, the first planned adrenalectomy was performed by Perry Sargent [3]. The first flank approach for a pheochromocytoma was performed by Charles Mayo in 1927 [4]. Most early adrenal surgeries were conducted to remove large tumors, and the incisions were essentially similar to those made for renal surgery. In adrenal surgery, the open method was the only surgical option until 1992 [5].

The first laparoscopic adrenalectomy (LA) using a lateral transperitoneal approach was reported in 1992 [6]. Another method of LA using a posterior retroperitoneal approach was introduced in 1995 [7]. Currently, LA is considered the gold standard for the excision of small, benign functional adrenal tumors [8]. With the emergence of robotic technology, robot-assisted adrenal gland surgery was first reported in 1999, and the first robotic adrenalectomy (RA) was reported in 2001 [9,10]. RA has gradually become popular in many countries, and the proportion of RAs has gradually been increasing [11,12].

The da Vinci robotic surgical device facilitates the use of many advanced techniques including high-quality three-dimensional (3D) vision and intuitive controlled movement with seven degrees of freedom through the endowrist function. Surgeons can perform surgery more comfortably and delicately using this robotic device, which can lead to better surgical results than conventional endoscopic surgery [13]. Robotic surgery may be useful in adrenalectomy, which involves precise movements in a limited space. However, research on RA suffers from a lack of cases compared with surgery on other organs [14,15].

Our hospital started performing robotic surgery using da Vinci Xi in 2018, and RA has also been performed. This study evaluates the initial experiences with RA compared with conventional LA performed by a single endocrine surgeon.

2. Methods

2.1. Study Design

We analyzed the electronic medical records of patients who underwent either LA or RA at Inha University Hospital. The potential surgical candidates for transperitoneal RA were the same as those for LA, and RA was chosen when the patient agreed to undergo surgery using this technique despite the greater expense than the laparoscopic approach. We reviewed clinical information data including age, gender, body mass index (BMI), preoperative clinical diagnosis, tumor location and size, type of combined operation, final pathology diagnosis, operation time, estimated blood loss volume, postoperative pain, postoperative length of hospital stay, complications, and surgical cost. All surgeries were recorded, and the videos were reviewed. We captured important scenes and calculated the actual surgery time based on the surgery videos. In the case of combined surgery, the operation time and cost of surgery were only evaluated for the adrenalectomy. The cost of the surgery was calculated as the fees for surgery only, excluding extra expenses including hospitalization and other expenses.

2.2. Patients

From October 2018 to March 2022, 62 adrenalectomies were performed by a single endocrine surgeon (JW Yi) at Inha University Hospital, Incheon, South Korea. All patients underwent adrenalectomy using a transperineal approach. Two groups were formed: the LA group and the RA group. We excluded three cases of open conversion (only occurred in LA) and three cases of initial open surgery. A total of 56 patients (24 patients in the RA group and 32 in the LA group) were included in the analysis.

2.3. Statistics and Ethical Considerations

All statistical analyses were performed using IBM SPSS Statistics 28 (IBM Corporation, Armonk, NY, USA). Continuous variables are presented using the mean $\pm$ standard deviation. Unpaired t-tests were used to compare the means. The chi-square test or Fisher's exact test was applied to the cross-table analysis, depending on the sample size.

The ethics of this study were approved by the institutional review board of the author's institution (INHAUH 2022-05-015).

2.4. Operative Procedure for RA

LA was performed via a traditional lateral transperitoneal approach, as previously reported elsewhere [16]. Our procedure for transperitoneal RA is described below.

Under general anesthesia, the patient was placed in the lateral decubitus position. The patient's bed was bent at an angle of 80° to expose the side as much as possible. Figure 1 shows the trocar placement in RA. For the right side, three robotic arms and an additional 5 mm port were required for liver mobilization, as shown in Figure 1A. For the left side, only three robotic arms were needed, as shown in Figure 1B. For effective movement of the robotic endowrist in the abdominal cavity, robot trocars were placed 3 cm apart from the subcostal margin to enable the unrestricted movement of the robotic arms. To prevent collision of the robot arms, a distance of approximately 5 cm was maintained between the trocars. We used three types of robotic endowrist instruments: prograsp forceps, long bipolar, and vessel sealer extend.

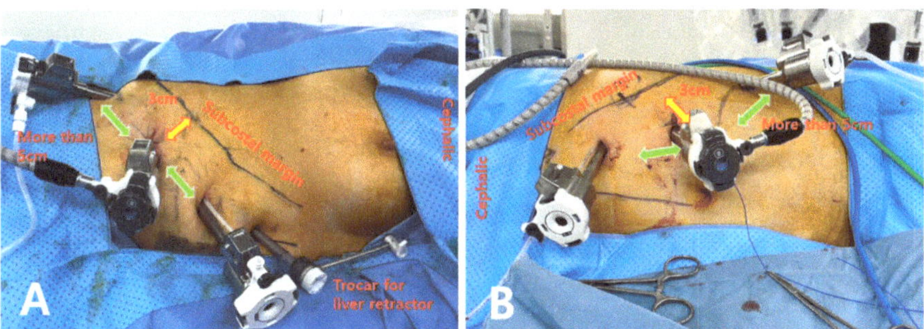

Figure 1. Trocar sites for robotic adrenalectomy. (**A**) Right adrenalectomy. (**B**) Left adrenalectomy.

The steps for right RA are shown in Figure 2. The right triangular ligament was detached to sufficiently mobilize the right liver (Figure 2A). Then, a snake retractor was inserted, the mobilized right liver pulled up, and the location of the inferior vena cava (IVC) was identified (Figure 2B). After opening Gerota's fascia, the upper pole of right kidney was identified. After dissecting upward along the superior pole of the right kidney, the right adrenal gland was identified (Figure 2C). The medial side of the adrenal gland was carefully dissected to find the adrenal vein (Figure 2D), and it was divided after hem-o-lock ligation (Figure 2E). The adrenal resection was terminated by completing the peripheral adrenal dissection (Figure 2F).

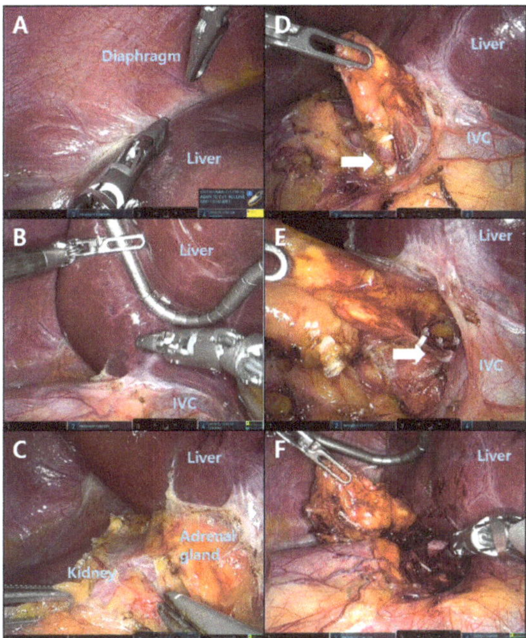

Figure 2. Procedure for robotic right adrenalectomy. (**A**) Liver mobilization. (**B**) Liver retraction and inferior vena cava identification. (**C**) Gerota's fascia dissection and adrenal gland identification. (**D**) Adrenal vein identification. (**E**). Clipping of the adrenal vein. (**F**) Adrenalectomy completion.

The steps for left RA are shown in Figure 3. The left colon was lowered by dissecting around the splenic flexure (Figure 3A). Then, the spleen was pulled downward by dissecting the splenocolic and splenophrenic ligaments (Figure 3B). After dissection around the distal pancreas and Gerota's fascia, the location of the adrenal gland was revealed (Figure 3C). The superior pole of the left kidney was exposed, and careful dissection was performed around the adrenal gland (Figure 3D). After identifying the adrenal vein, it was ligated by hem-o-lock and cut (Figure 3E). The remaining soft tissue around the adrenal gland was detached, and the adrenalectomy was completed (Figure 3F).

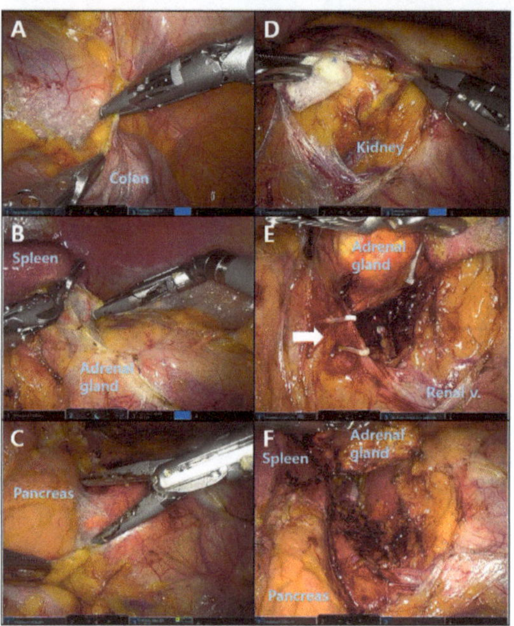

Figure 3. Procedure for robotic left adrenalectomy. (**A**) Left colon mobilization. (**B**) Spleen mobilization. (**C**) Pancreas and Gerota's fascia dissection. (**D**) Exposure of the left kidney. (**E**) Adrenal vein identification and ligation. (**F**) Adrenalectomy completion.

3. Results

The clinical characteristics of the RA and LA groups are summarized in Table 1. The mean age in the RA group was significantly lower than that in the LA group (48.6 ± 9.7 years vs. 56.1 ± 11.4 years, $p = 0.013$). Gender and BMI were not different between the two groups. The clinical diagnoses included primary aldosteronism, Cushing syndrome, and the increasing size of a non-functioning adenoma. There were no significant differences in the tumor location and size between the two groups. The RA group had four combined operation cases with a robotic approach including cholecystectomy, partial nephrectomy, and total hysterectomy with bilateral salpingo-oophorectomy. The LA group had one combined operation case, i.e., laparoscopic cholecystectomy.

Table 2 summarizes the surgical and clinical outcomes of the RA and LA groups. The RA group exhibited a shorter operation time than the LA group (76.1 ± 28.2 min vs. 118.0 ± 54.3 min, $p < 0.001$). There were no significant differences in the estimated blood loss, hospital stay after surgery, and visual analogue pain scale on postoperative days 1 and 2 between the two groups. There were no postoperative complications in the RA group, while a port-site hernia occurred in a single case in the LA group. The cost of surgery was significantly higher in the RA group than in the LA group (5288 US dollars vs. 441.5 ± 136.8 US dollars, $p < 0.001$). The pathologic diagnoses included adrenal cortical adenoma, pheochromocytoma/paraganglioma, and myelolipoma. There was no significant difference in the pathologic diagnosis between the two groups.

Table 1. Clinical characteristics of patients undergoing robotic adrenalectomy and laparoscopic adrenalectomy.

Variable	Total ($n = 56$)	RA ($n = 24$)	LA ($n = 32$)	p-Value
Age (years, mean $\pm$ SD)	52.9 $\pm$ 11.2 (25–76)	48.6 $\pm$ 9.7 (29–71)	56.1 $\pm$ 11.4 (25–76)	0.013
Gender				
Male	28 (50%)	12 (21%)	16 (29%)	1.000
Female	28 (50%)	12 (21%)	16 (29%)	
BMI [a] (kg/m^2, mean $\pm$ SD)	25.9 $\pm$ 4.4 (17.3–35.0)	26.6 $\pm$ 4.9 (18.2–35.0)	25.4 $\pm$ 4.1 (17.3–32.7)	0.344
Clinical diagnosis				
Primary aldosteronism	16 (28.6%)	9 (37.5%)	7 (21.9%)	
Cushing's syndrome	14 (25.0%)	6 (25.0%)	8 (25%)	
Non-functioning Adenoma	12 (21.4%)	3 (12.5%)	9 (28.1%)	
PPGLs [b]	9 (16.1%)	5 (20.8%)	4 (12.5%)	
r/o Metastatic mass	3 (5.4%)	1 (4.2%)	2 (6.25%)	
Adrenal cortical cancer	2 (3.6%)	0 (0.0%)	2 (6.25%)	
Tumor location				
Right	27 (48%)	14 (25%)	13 (23%)	0.196
Left	29 (52%)	10 (18%)	19 (34%)	
Tumor size (cm, mean $\pm$ SD)	3.5 $\pm$ 2.2 (1.3–9.6)	3.1 $\pm$ 1.8 (1.3–8.0)	3.8 $\pm$ 2.4 (1.3–9.6)	0.269
Combined operation	3			
Cholecystectomy	2	2	1	
Partial nephrectomy	1	2	0	
TH + BSO [c]	3	1	0	

[a] Body mass index, [b] Pheochromocytoma/paraganglioma [c] Total hysterectomy + bilateral salpingo-oophorectomy.

Table 2. Surgical and clinical outcomes of robotic adrenalectomy and laparoscopic adrenalectomy.

Variable	Total ($n = 56$)	RA ($n = 24$)	LA ($n = 32$)	p-Value
Operation time (minute, mean $\pm$ SD)	100.0 $\pm$ 48.6 (46–295)	76.1 $\pm$ 28.2 (46–140)	118.0 $\pm$ 54.3 (60–295)	<0.001
Estimated blood loss (mL, mean $\pm$ SD)	152.1 $\pm$ 132.7 (0–400)	164.6 $\pm$ 144.8 (0–400)	142.8 $\pm$ 124.5 (0–400)	0.548
Hospital stay after operation (days, mean $\pm$ SD)	4.0 $\pm$ 1.9 (2–9)	4.3 $\pm$ 2.1 (2–9)	3.8 $\pm$ 1.7 (2–9)	0.326
Visual analog scale				
Postoperative day 1 (0–10, mean $\pm$ SD)	2.9 $\pm$ 0.6 (0–5)	2.9 $\pm$ 0.9 (0–5)	2.9 $\pm$ 0.4 (2–4)	0.737
Postoperative day 2 (0–10, mean $\pm$ SD)	2.5 $\pm$ 0.7 (0–3)	2.5 $\pm$ 0.8 (0–3)	2.5 $\pm$ 0.7 (1–3)	0.873
Complication				
Port-site hernia	1 (1.8%)	0 (0.0%)	1 (3.1%)	0.382
Cost of surgery (US dollar [c], mean $\pm$ SD)	2528.8 $\pm$ 2422.5 (352.8–5288.5)	5288.5	441.5 $\pm$ 136.8 (352.8–1167.1)	<0.001
Pathologic diagnosis				
Adrenal cortical adenoma	35 (62.5%)	16 (66.7%)	19 (59.4%)	
PPGLs [a]	6 (10.7%)	5 (20.8%)	1 (3.1%)	
Myelolipoma	5 (8.9%)	2 (3.6%)	3 (9.4%)	
Adrenal cortical cancer	2 (3.6%)	0 (0.0%)	2 (6.3%)	
Others [b]	8 (14.3%)	1 (4.2%)	7 (21.9%)	

[a] Pheochromocytoma/paraganglioma, [b] Adenomatoid tumor, Cortical nodular hyperplasia, Ganglioneuroma, Lymphangioma, Macronodular hyperplasia, Malignant pheochromocytoma, Metastatic Hepatocellular cell, Pseudocys, [c] Exchange rate as of 17 April 2022.

Table 3 summarizes the results of other studies in which RAs were performed. In our study, the tumor size was similar to that in previous studies, but the operation time was reported to be shorter than that in previous studies.

Table 3. Case series for robotic adrenalectomy.

Year		Study Design	Patient (*n*)	Tumor Size (cm)	Operation Time (min)
2006	Winter [17]	Case series	30	2.4	185
2008	Brunaaud [18]	Case series	100	2.9	171
2011	Giulianotti [19]	Case series	42	5.5	118
2011	Nordenström [20]	Case series	100	5.3	113
2012	Agcaoglu [21]	Comparative	31	3.1	163.2
2012	D'Annibale [22]	Case series	30	5.1	200
2013	Aksoy [23]	Comparative	42	4.0	186
2013	Aliyev [24]	Comparative	26	-	149
2014	Brandao [25]	Comparative	30	3	120
2016	Lee [26]	Case series	33	-	234
2016	Morelli [27]	Comparative	41	-	177
2019	Greilsamer [28]	Case series	303	3.6	89
2019	Kim [29]	Comparative	61	3.7	138
2020	Carmela [30]	Comparative	12	-	93.3
2020	Fu [31]	Comparative	19	8	166.3
2020	Changwei [32]	Comparative	87	4.7	136.1
2020	Ozdemir [33]	Case series	111	3.9	135.4
2020	Fang [34]	Comparative	41	6.2	210.4
2021	Piccoli [35]	Comparative	76	4.0	100.3
2022	Knežević [36]	Case series	12	-	165.1
2022	Erdemir [37]	Case series	30	8.3	194.9
2022	Al-Thani [38]	Comparative	76	4.8	174

4. Discussion

Compared with conventional open surgery, minimally invasive surgery (MIS) has many advantages: it is not only cosmetically superior but also results in less postoperative pain, a shorter hospital stay, and favorable oncologic outcomes in cancer surgery [39–41]. MIS has been mainly performed using a laparoscopic approach. With the development of robotic surgical systems, MIS using a robotic approach has been established, and many clinical studies have been conducted [42].

LA allows for better access to adrenal glands than open surgery. The pneumoperitoneum enables a wider surgical space into which other organs can easily be retracted in a downward direction. Dissection around the adrenal glands is also easier because of the fine and straightforward laparoscopic instruments. The adrenal vessels can be easily ligated using a laparoscopic clip applier compared with open adrenalectomy [43,44]. The size of the incision is very small, and only three or four incisions are required for trocar placement compared with open methods. Given these advantages, LA is considered the standard method [8].

The first robotic adrenal surgery was reported in 1999 [9]. Currently, several clinical experiences of robotic adrenalectomy are being reported in various hospitals as described in Table 3. Some studies that compared RA to the laparoscopic approach suggested that there are no specific advantages of RA [21,23,24,27,29,32,34,35]. However, we found that the operation time was significantly shorter in the RA group, as described in Table 1. Furthermore, there were no postoperative complications in the RA group. We suggest that the reason for these results may be that the surgeon who performed RA at our hospital had considerable experience in robotic surgery, i.e., had performed more than 500 robotic thyroid surgeries. There was no need for adaptation or a learning curve to perform robotic surgery. In addition, he had sufficient LA experience.

In our study, there was one case of a port-site hernia in the LA group. It occurred two months after surgery through the 12 mm port site and was corrected by surgical treatment. This may have been due to surgical failure to close the 12 mm port site. As 5 mm and 8 mm trocars are used in robot adrenalectomy, it may be helpful to prevent port-site hernia. We performed two cases of right RA with cholecystectomy. The lateral decubitus position is used in adrenalectomy, whereas the supine position is used in cholecystectomy. We safely performed both surgeries in a single stage using the lateral decubitus position without changing the position. We believe that right RA and cholecystectomy can be performed safely in a single stage using the lateral decubitus position without any special position changes.

The advantages of the robotic system in an adrenalectomy are as follows. First, there is considerable extracorporeal and intracorporeal fighting in LA because the surgeon and assistant are too close during surgery. In contrast, RA does not involve such collision, and more free surgical movement is possible. Second, more and precise free movement is possible by using the endowrist of the robot arm than by using the laparoscopic device. This is very convenient for liver mobilization and enables increased mobilization. In LA, there is also a risk of surrounding structure injury because the angle of the laparoscopic arm is not parallel to the IVC. The angulation of the robotic arm makes this possible and reduces this risk. Third, the 3D augmented view of the robotic system provides a very clear and accurate view to the operator, and the operator can directly control it to obtain the desired view.

The disadvantages of RA are as follows: First, it does not provide tactile sense from the instrument to the surgeon's hand. This makes it difficult for inexperienced surgeons to distinguish tissues or organs. Second, RA is more than five times as expensive as LA. Many people have a personal medical insurance system in South Korea, and patients pay only 0–20% of the surgical cost. In this case, there is no difference between robotic surgery and laparoscopic surgery. RA showed some advantages compared to LA in our study. RA can be recommended to patients with personal medical insurance. The final concern is proper training for surgeons. Although robotic surgery is popular worldwide, there are few institutions that perform RA on a large scale. This makes it difficult for surgeons to learn about robotic adrenal surgery.

One limitation of our study is that RA is not yet a widely used surgical technique in South Korea; thus, there are not enough surgical cases. There are not many indications for adrenalectomy, and there are not many institutions that perform adrenalectomy in South Korea. Nevertheless, it is meaningful to analyze 56 cases over approximately two years. As more cases are accumulated in the future, additional analyses will be conducted.

5. Conclusions

The transperitoneal RA is a promising surgical method for adrenalectomy. RA has many advantages over conventional LA. The operation time of RA was shorter than that of LA, and no significant complications occurred.

Several studies have reported their experiences with adrenalectomy using a single-port robot system [17,44]. We will adopt a new single-port robot system in the near future and evaluate the advantages and disadvantages of this new robot system.

Author Contributions: Conceptualization: J.W.Y.; Data curation: J.W.Y.; Formal analysis: J.S.L.; Investigation: J.S.L.; Methodology: J.W.Y.; Project administration: J.W.Y.; Resources: J.W.Y.; Supervision: J.W.Y.; Validation: Y.S.C.; Writing—original draft: Y.S.C.; Writing—review and editing: J.W.Y. All authors have read and agreed to the published version of the manuscript.

Funding: This research received no external funding.

Institutional Review Board Statement: This study was approved by the institutional review board of the author's institution (INHAUH 2022-05-015).

Informed Consent Statement: Patient consent was waived due to retrospective study with medical record.

Data Availability Statement: No new data were created or analyzed in this study. Data sharing is not applicable to this article.

Acknowledgments: This study was supported by a research grant from Inha University Hospital.

Conflicts of Interest: The authors declare no conflict of interest.

References

1. Shah, M.H.; Goldner, W.S.; Benson, A.B.; Bergsland, E.; Blaszkowsky, L.S.; Brock, P.; Chan, J.; Das, S.; Dickson, P.V.; Fanta, P. Neuroendocrine and adrenal tumors, version 2.2021, NCCN Clinical Practice Guidelines in Oncology. *J. Natl. Compr. Cancer Netw.* **2021**, *19*, 839–868. [CrossRef]
2. Thornton, J. Abdominal nephrectomy for large sarcoma of the left suprarenal capsule: Recovery. *Trans. Clin. Soc. Lond.* **1890**, *23*, 150–153.
3. Hughes, S.; Lynn, J. Surgical anatomy and surgery of the adrenal glands. In *Surgical Endocrinology*; Elsevier: Amsterdam, The Netherlands, 1993; pp. 458–467.
4. Mayo, C.H. Paroxysmal hypertension with tumor of retroperitoneal nerve: Report of case. *J. Am. Med. Assoc.* **1927**, *89*, 1047–1050. [CrossRef]
5. Harris, D.A.; Wheeler, M.H. History of adrenal surgery. In *Adrenal Glands*; Springer: Berlin/Heidelberg, Germany, 2005; pp. 1–6.
6. Gagner, M. Laparoscopic adrenalectomy in Cushing's syndrome and pheochromocytoma. *N. Engl. J. Med.* **1992**, *327*, 1033. [PubMed]
7. Mercan, S.; Seven, R.; Ozarmagan, S.; Tezelman, S. Endoscopic retroperitoneal adrenalectomy. *Surgery* **1995**, *118*, 1071–1076. [CrossRef] [PubMed]
8. Smith, C.D.; Weber, C.J.; Amerson, J.R. Laparoscopic adrenalectomy: New gold standard. *World J. Surg.* **1999**, *23*, 389. [CrossRef]
9. Piazza, L.; Caragliano, P.; Scardilli, M.; Sgroi, A.; Marino, G.; Giannone, G. Laparoscopic robot-assisted right adrenalectomy and left ovariectomy. *Chir. Ital.* **1999**, *51*, 465–466.
10. Horgan, S.; Vanuno, D. Robots in laparoscopic surgery. *J. Laparoendosc. Adv. Surg. Tech.* **2001**, *11*, 415–419. [CrossRef]
11. Grogan, R.H. Current status of robotic adrenalectomy in the United States. *Gland Surg.* **2020**, *9*, 840. [CrossRef]
12. Makay, O.; Erol, V.; Ozdemir, M. Robotic adrenalectomy. *Gland Surg.* **2019**, *8*, S10. [CrossRef]
13. Palep, J.H. Robotic assisted minimally invasive surgery. *J. Minimal Access Surg.* **2009**, *5*, 1. [CrossRef] [PubMed]
14. Economopoulos, K.P.; Mylonas, K.S.; Stamou, A.A.; Theocharidis, V.; Sergentanis, T.N.; Psaltopoulou, T.; Richards, M.L. Laparoscopic versus robotic adrenalectomy: A comprehensive meta-analysis. *Int. J. Surg.* **2017**, *38*, 95–104. [CrossRef] [PubMed]
15. Nomine-Criqui, C.; Demarquet, L.; Schweitzer, M.L.; Klein, M.; Brunaud, L.; Bihain, F. Robotic adrenalectomy: When and how? *Gland Surg.* **2020**, *9*, S166–S172. [CrossRef]
16. Chai, Y.J.; Yu, H.W.; Song, R.-Y.; Kim, S.-J.; Choi, J.Y.; Lee, K.E. Lateral transperitoneal adrenalectomy versus posterior retroperitoneoscopic adrenalectomy for benign adrenal gland disease: Randomized controlled trial at a single tertiary medical center. *Ann. Surg.* **2019**, *269*, 842–848. [CrossRef] [PubMed]
17. Winter, J.; Talamini, M.; Stanfield, C.; Chang, D.; Hundt, J.; Dackiw, A.; Campbell, K.; Schulick, R. Thirty robotic adrenalectomies. *Surg. Endosc. Other Interv. Tech.* **2006**, *20*, 119–124. [CrossRef] [PubMed]
18. Brunaud, L.; Ayav, A.; Zarnegar, R.; Rouers, A.; Klein, M.; Boissel, P.; Bresler, L. Prospective evaluation of 100 robotic-assisted unilateral adrenalectomies. *Surgery* **2008**, *144*, 995–1001. [CrossRef] [PubMed]
19. Giulianotti, P.; Buchs, N.; Addeo, P.; Bianco, F.; Ayloo, S.; Caravaglios, G.; Coratti, A. Robot-assisted adrenalectomy: A technical option for the surgeon? *Int. J. Med. Robot. Comput. Assist. Surg.* **2011**, *7*, 27–32. [CrossRef]
20. Nordenström, E.; Westerdahl, J.; Hallgrimsson, P.; Bergenfelz, A. A prospective study of 100 robotically assisted laparoscopic adrenalectomies. *J. Robot. Surg.* **2011**, *5*, 127–131. [CrossRef]
21. Agcaoglu, O.; Aliyev, S.; Karabulut, K.; Siperstein, A.; Berber, E. Robotic vs laparoscopic posterior retroperitoneal adrenalectomy. *Arch. Surg.* **2012**, *147*, 272–275. [CrossRef]
22. D'Annibale, A.; Lucandri, G.; Monsellato, I.; De Angelis, M.; Pernazza, G.; Alfano, G.; Mazzocchi, P.; Pende, V. Robotic adrenalectomy: Technical aspects, early results and learning curve. *Int. J. Med. Robot. Comput. Assist. Surg.* **2012**, *8*, 483–490. [CrossRef]
23. Aksoy, E.; Taskin, H.E.; Aliyev, S.; Mitchell, J.; Siperstein, A.; Berber, E. Robotic versus laparoscopic adrenalectomy in obese patients. *Surg. Endosc.* **2013**, *27*, 1233–1236. [CrossRef] [PubMed]
24. Aliyev, S.; Karabulut, K.; Agcaoglu, O.; Wolf, K.; Mitchell, J.; Siperstein, A.; Berber, E. Robotic versus laparoscopic adrenalectomy for pheochromocytoma. *Ann. Surg. Oncol.* **2013**, *20*, 4190–4194. [CrossRef] [PubMed]
25. Brandao, L.F.; Autorino, R.; Zargar, H.; Krishnan, J.; Laydner, H.; Akca, O.; Mir, M.C.; Samarasekera, D.; Stein, R.; Kaouk, J. Robot-assisted laparoscopic adrenalectomy: Step-by-step technique and comparative outcomes. *Eur. Urol.* **2014**, *66*, 898–905. [CrossRef] [PubMed]
26. Lee, G.S.; Arghami, A.; Dy, B.M.; McKenzie, T.J.; Thompson, G.B.; Richards, M.L. Robotic single-site adrenalectomy. *Surg. En-dosc.* **2016**, *30*, 3351–3356. [CrossRef]
27. Morelli, L.; Tartaglia, D.; Bronzoni, J.; Palmeri, M.; Guadagni, S.; Di Franco, G.; Gennai, A.; Bianchini, M.; Bastiani, L.; Moglia, A. Robotic assisted versus pure laparoscopic surgery of the adrenal glands: A case-control study comparing surgical techniques. *Langenbeck's Arch. Surg.* **2016**, *401*, 999–1006. [CrossRef] [PubMed]

28. Greilsamer, T.; Nomine-Criqui, C.; Thy, M.; Ullmann, T.; Zarnegar, R.; Bresler, L.; Brunaud, L. Robotic-assisted unilateral adrenalectomy: Risk factors for perioperative complications in 303 consecutive patients. *Surg. Endosc.* **2019**, *33*, 802–810. [CrossRef]
29. Kim, W.W.; Lee, Y.M.; Chung, K.W.; Hong, S.J.; Sung, T.Y. Comparison of robotic posterior retroperitoneal adrenalectomy over laparoscopic posterior retroperitoneal adrenalectomy: A single tertiary center experience. *Int. J. Endocrinol.* **2019**, *2019*, 9012910. [CrossRef]
30. De Crea, C.; Arcuri, G.; Pennestrì, F.; Paolantonio, C.; Bellantone, R.; Raffaelli, M. Robotic adrenalectomy: Evaluation of cost-effectiveness. *Gland Surg.* **2020**, *9*, 831–839. [CrossRef]
31. Fu, S.-Q.; Zhuang, C.-S.; Yang, X.-R.; Xie, W.-J.; Gong, B.-B.; Liu, Y.-F.; Liu, J.; Sun, T.; Ma, M. Comparison of robot-assisted retroperitoneal laparoscopic adrenalectomy versus retroperitoneal laparoscopic adrenalectomy for large pheochromocytoma: A single-centre retrospective study. *BMC Surg.* **2020**, *20*, 227. [CrossRef]
32. Ji, C.; Lu, Q.; Chen, W.; Zhang, F.; Ji, H.; Zhang, S.; Zhao, X.; Li, X.; Zhang, G.; Guo, H. Retrospective comparison of three minimally invasive approaches for adrenal tumors: Perioperative outcomes of transperitoneal laparoscopic, retroperitoneal laparoscopic and robot-assisted laparoscopic adrenalectomy. *BMC Urol.* **2020**, *20*, 66. [CrossRef]
33. Ozdemir, M.; Dural, A.C.; Sahbaz, N.A.; Akarsu, C.; Uc, C.; Sertoz, B.; Alis, H.; Makay, O. Robotic transperitoneal adrenalectomy from inception to ingenuity: The perspective on two high-volume endocrine surgery centers. *Gland Surg.* **2020**, *9*, 815–825. [CrossRef] [PubMed]
34. Fang, A.M.; Rosen, J.; Saidian, A.; Bae, S.; Tanno, F.Y.; Chambo, J.L.; Bloom, J.; Gordetsky, J.; Srougi, V.; Phillips, J.; et al. Perioperative outcomes of laparoscopic, robotic, and open approaches to pheochromocytoma. *J. Robot. Surg.* **2020**, *14*, 849–854. [CrossRef] [PubMed]
35. Agcaoglu, O.; Karahan, S.N.; Tufekci, T.; Tezelman, S. Single-incision robotic adrenalectomy (SIRA): The future of adrenal sur-gery? *Gland Surg.* **2020**, *9*, 853–858. [CrossRef] [PubMed]
36. Knežević, N.; Penezić, L.; Kuliš, T.; Zekulić, T.; Saić, H.; Hudolin, T.; Kaštelan, Ž. Senhance robot-assisted adrenalectomy: A case series. *Croat. Med. J.* **2022**, *63*, 197–201. [CrossRef]
37. Erdemir, A.; Rasa, K. Robotic adrenalectomy: An initial experience in a Turkish regional hospital. *Front. Surg.* **2022**, *9*, 847472. [CrossRef] [PubMed]
38. Al-Thani, H.; Al-Thani, N.; Al-Sulaiti, M.; Tabeb, A.; Asim, M.; El-Menyar, A. A descriptive comparative analysis of the surgical management of adrenal tumors: The open, robotic, and laparoscopic approaches. *Front. Surg.* **2022**, *9*, 848565. [CrossRef] [PubMed]
39. Lee, K.E.; Rao, J.; Youn, Y.-K. Endoscopic thyroidectomy with the da Vinci robot system using the bilateral axillary breast ap-proach (BABA) technique: Our initial experience. *Surg. Laparosc. Endosc. Percutaneous Tech.* **2009**, *19*, e71–e75. [CrossRef]
40. Yang, H.K.; Suh, Y.S.; Lee, H.J. Minimally invasive approaches for gastric cancer—Korean experience. *J. Surg. Oncol.* **2013**, *107*, 277–281. [CrossRef]
41. Mendoza, A.S., III; Han, H.S.; Yoon, Y.S.; Cho, J.Y.; Choi, Y. Laparoscopy-assisted pancreaticoduodenectomy as minimally in-vasive surgery for periampullary tumors: A comparison of short-term clinical outcomes of laparoscopy-assisted pancreatico-duodenectomy and open pancreaticoduodenectomy. *J. Hepato-Biliary-Pancreat. Sci.* **2015**, *22*, 819–824. [CrossRef]
42. Barbash, G.I. New technology and health care costs—The case of robot-assisted surgery. *N. Engl. J. Med.* **2010**, *363*, 701. [CrossRef]
43. Yeganeh, H. An analysis of emerging trends and transformations in global healthcare. *Int. J. Health Gov.* **2019**, *24*, 169–180. [CrossRef]
44. Park, J.H.; Walz, M.K.; Kang, S.-W.; Jeong, J.-J.; Nam, K.-H.; Chang, H.-S.; Chung, W.-Y.; Park, C.-S. Robot-assisted posterior retroperitoneoscopic adrenalectomy: Single port access. *J. Korean Surg. Soc.* **2011**, *81*, S21–S24. [CrossRef] [PubMed]

medicina

Article

Diagnostic Accuracy of Ultrasound and Fine-Needle Aspiration Cytology in Thyroid Malignancy

Maria Boudina [1], Michael Katsamakas [2], Angeliki Chorti [3], Panagiotis Panousis [2], Eleni Tzitzili [2], Georgios Tzikos [3], Alexandra Chrisoulidou [1], Rosalia Valeri [4], Aris Ioannidis [3] and Theodossis Papavramidis [3,*]

1 Department of Endocrinology, Theageneio Cancer Hospital, 54636 Thessaloniki, Greece
2 Department of Surgery, Theageneio Cancer Hospital, 54636 Thessaloniki, Greece; panagiotis.panousis@gmail.com (P.P.)
3 1st Propaedeutic Department of Surgery, AHEPA University Hospital, Aristotle University, 54636 Thessaloniki, Greece
4 Department of Pathology, Theageneio Cancer Hospital, 54636 Thessaloniki, Greece
* Correspondence: papavramidis@hotmail.com or tpapavr@auth.gr; Tel.: +30-6944536972

Abstract: *Introduction*: Thyroid nodule incidence is increasing due to the widespread application of ultrasonography. Fine-needle aspiration cytology is widely applied for the detection of malignancies. The aim of this study was to evaluate the predictive value of ultrasonography in thyroid cancer. *Methods*: This retrospective study included patients that underwent total thyroidectomy for benign thyroid disease or well-differentiated thyroid carcinoma from January 2017 to December 2022. The study population was divided into groups: the well-differentiated thyroid cancer group and the control group with benign histopathological reports. *Results*: In total, 192 patients were enrolled in our study; 159 patients were included in the well-differentiated thyroid cancer group and 33 patients in the control group. Statistical analysis demonstrated that ultrasonographic findings such as microcalcifications (90.4%), hypoechogenicity (89.3%), irregular margins (92.2%) and taller-than-wide shape (90.5%) were correlated to malignancy ($p < 0.001$). Uni- and multivariate analysis revealed that both US score (OR: 2.177; $p < 0.001$) and Bethesda System (OR: 1.875; $p = 0.002$) could predict malignancies. In terms of diagnostic accuracy, the US score displayed higher sensitivity (64.2% vs. 33.3%) and better negative predictive value (34.5% vs. 24.4%) than the Bethesda score, while both scoring systems displayed comparable specificities (90.9% vs. 100%) and positive predictive values (97.1% vs. 100%). *Discussion*: The malignant potential of thyroid nodules is a crucial subject, leading the decision for surgery. Ultrasonography and fine-needle aspiration cytology are pivotal examinations in the diagnostic process, with ultrasonography demonstrating better negative predictive value.

Keywords: thyroid malignancy; fine-needle aspiration; ultrasound; thyroid nodule

Citation: Boudina, M.; Katsamakas, M.; Chorti, A.; Panousis, P.; Tzitzili, E.; Tzikos, G.; Chrisoulidou, A.; Valeri, R.; Ioannidis, A.; Papavramidis, T. Diagnostic Accuracy of Ultrasound and Fine-Needle Aspiration Cytology in Thyroid Malignancy. *Medicina* **2024**, *60*, 722. https://doi.org/10.3390/medicina60050722

Academic Editor: Jin Wook Yi

Received: 3 April 2024
Revised: 21 April 2024
Accepted: 25 April 2024
Published: 26 April 2024

1. Introduction

The incidence of thyroid nodules (TNs) has been increasing over the last years due to the wide use of ultrasonography (US) and other imaging tests. Reportedly, TNs can be found in 2–6% of the population with palpation, in 19–35% on ultrasound and in 8–65% in autopsy series [1]. Since the incidence of thyroid malignancy has been reported to 5.4% for men and 6.5% for women, only a small fraction of those nodules prove to be malignant [2]. Fine Needle Aspiration Cytology (FNAc) is a widely accepted method for the evaluation of TNs and the detection of malignancy with reported accuracy rates varying and exceeding 90% in some reports [3]. FNAc has indisputably contributed to the decrease in the number of unnecessary thyroid surgeries and to the increase of the preoperative diagnosis of malignant thyroid lesions. The 2023 Bethesda System for Reporting Thyroid Cytopathology recommends six reporting categories: (i) nondiagnostic; (ii) benign; (iii) atypia of undetermined significance (AUS); (iv) follicular neoplasm; (v) suspicious for malignancy (SFM); and (vi) malignant [4]. Factors that can influence FNAc results are a possibly inaccessible position of the nodule, operator experience, nodule size and composition, as well as experience in cytology interpretation.

Many authors have advocated the importance of establishing US criteria for diagnosing thyroid malignancy, which will serve as an additional source of information for the clinician facing a diagnostic dilemma. Certain US characteristics have previously been strongly correlated with malignancy, such as hypoechogenicity, irregular margins, taller-than-wide shape and microcalcifications. In 2015 American Thyroid Association published Management Guidelines for adults with TNs and differentiated thyroid carcinoma, associating sonographic patterns of TNs with the risk of malignancy [5]. In order to assess the risk of malignancy certain sonographic criteria are used. According to ATA, TNs are discriminated as following: TIRADS 1: normal, TIRADS 2: benign conditions, TIRADS 3: probably benign nodules (<5%), TIRADS 4: suspicious nodules (5–80%), TIRADS 5: probably malignant (>80%), TIRADS 6: Biopsy proven malignancy. Risk stratification criteria were conducted by the Korean Thyroid Association (K-TIRADS) [6]. The European Thyroid Association designed another reporting system for ultrasound assessment, EU TIRADS. There are five EU-TIRADS categories: EU-TIRADS 1: normal thyroid lesions, EU-TIRADS 2: benign lesions such us cysts, EU-TIRADS 3: low risk (2–4%) lesions as for isoechoic/hyperechoic nodules with smooth margins, EU-TIRADS 4: intermediate risk (6–17%) lesions such as ovoid, mildly hypoechoic nodules with smooth margins, EU-TIRADS 5: high risk (26–87%) lesions for nodules with suspicious characteristics such as irregular shape or margins, micro-calcifications, taller than wide morphology and markedly hypoechoic solid lesions [7]. Finally, the British Thyroid Association (BTA) classified TNs into 5 categories; U1 = normal thyroid gland, U2 = benign TN, U3 = intermediate/equivocal TN, U4 = suspicious TN, and U5 = malignant TN [8].

The aim of this study was to evaluate the predictive value of ultrasonographic features for malignant TNs and to assess the diagnostic performance of these features in thyroid cancer patients.

2. Materials and Methods

2.1. Study Population

Approval for this retrospective study was obtained from the institutional review board of our hospital (Scientific Council of Theageneio Cancer Hospital, Thessaloniki, Greece, 2661/22 February 2024), and informed consent was obtained from all patients. From January 2017 to December 2022, all patients aged >18 years old who underwent total thyroidectomy in our institution with a histopathology report of a benign thyroid gland or well-differentiated thyroid carcinoma were eligible for the study. Exclusion criteria were insufficient data on the preoperative evaluation of the patients, previous thyroid surgery and a pathology or cytology report of other types of malignancies apart from well-differentiated thyroid cancer.

The studied population was divided into two groups, the malignant group (WDTC-group) which included patients with a histopathological diagnosis of well-differentiated thyroid cancer following thyroidectomy, and the control group (NC-group) consisting of patients with a benign histopathology report. All patients included in this study were preoperatively submitted to a head and neck ultrasound and an ultrasound-guided FNAc (US-FNAc) of the suspicious nodules.

2.2. Ultrasonography and US-FNAc

A complete thyroid and neck ultrasound was acquired preoperatively from all patients. The examination was conducted by experienced radiologists (>than 5 years of experience). A US scoring system was implemented in accordance with the 2015 ATA Guidelines, associating thyroid nodules with the risk of malignancy based on their sonographic pattern. The thyroid nodules were assessed regarding the following ultrasonographic features: size, microcalcifications, increased vascularity, hypoechogenicity, taller-than-wide shape, irregular margins, extrathyroidal extension and solid composition. Each of the aforementioned features was appointed one point in the scoring system, reported as US score. Regarding the size, one point was given when the size was more than 10 mm.

A non-aspiration technique using a 23-gauge needle attached to a 5 mL syringe was performed. The samples were evaluated by experienced cytopathologists that were blinded to the US findings. The results of the FNAc were classified into the following six categories according to the Bethesda System for Reporting Thyroid Cytology: (1) nondiagnostic or unsatisfactory (Bethesda System I), (2) benign (Bethesda System II), (3) atypia or follicular lesion of undetermined significance (AUS/FLUS) (Bethesda System III), (4) follicular neoplasm or suspicious for a follicular neoplasm (Bethesda System IV), (5) suspicious for malignancy (Bethesda System V) and (6) malignant (Bethesda System VI). In cases of nodules with both cystic and solid components, an FNAc of the solid part of the nodule was performed. When a multinodular goiter was present, both the largest nodule and the nodule with the most suspicious characteristics were aspirated.

2.3. Data and Statistical Analysis

Data were analyzed in the Statistical Package for the Social Sciences 25.0 (SPSS Inc., Chicago, IL, USA) and R Software (version 3.6.2). Relationships with a two-sided p-value of less than 0.05 were considered statistically significant. The reference standard for malignancy was the histopathology report. In cases in which both the largest nodule and the nodule with the most suspicious characteristics were aspirated and assessed, the data were analyzed cumulatively, like both nodules being independent. Continuous variables were demonstrated as means with standard deviation (SD) or as medians with interquartile ranges (IQRs), depending on normality having been assumed or not, respectively, while categorical variables were presented as frequencies with percentages (%). The Chi-square test (X^2) and Fisher's exact test were applied to investigate the malignancy rate in categorical variables for US features. Independent samples t-tests and the non-parametric test of Mann–Whitney were used to evaluate the relationship between continuous variables and malignancies. Univariate and multivariate logistic regression analysis was performed to predict the probability of cancer for both the US score and the Bethesda System. Gender and age at the time of surgery were both included in the final model. After the univariate logistic analysis of every possible factor had been performed, a multivariate logistic analysis was conducted and the final model was built. During univariate regression, a factor was included in the multivariate logistic regression model when it met a statistical significance of a p-value less than 0.20. The final model was built using a stepwise backward elimination method with a significance level of 0.05.

In addition, a receiver operating characteristic (ROC) curve was generated to calculate the optimal cut-off point of the US score for thyroid malignancy which was chosen based on the accompanying Youden's index; sensitivity and specificity were also measured. After the optimal cut-off points for NLR and PLR had been calculated, then the sample size was divided into two groups based on them and the mortality incidence was reassessed [9,10].

3. Results

3.1. Basic Characteristics

One hundred and ninety-two patients were included in the study. The WDTC group included 159 patients and the NC group included 33 patients. No statistically significant differences were observed in the demographic characteristics between the two groups. Of the total 192 patients included in this study, 162 were female (84.4%) and 30 were male (15.6%). The mean age at the time of the surgery was 52.2 (SD: 13.3) years. The majority of the patients in both groups (n = 154, 80.6%) presented with a multinodular goiter (Table 1).

Table 1. Basic characteristics of the study population.

	All Patients (*n* = 192)	WDTC Group (*n* = 159)	NC Group (*n* = 33)	*p*-Value
Female Gender (%)	162 (84.4%)	136 (84.0%)	26 (16.0%)	0.331
Mean age at surgery [y] (SD)	52.2 (13.3)	52.1 (13.8)	52.2 (11.0)	0.971
Preoperative median TSH [ng/dL] (IQR)	1.53 (1.60)	1.60 (1.71)	1.71 (1.24)	0.479
Multinodular goiter (%)	154 (80.6%)	125 (81.2%)	29 (18.8%)	0.247

3.2. US Features

In total, 246 nodules were evaluated. Nodule size ranged from 8.0 to 58.0 mm, with a median of 21.0 mm (IQR: 17.3 mm). Malignant nodules tended to be slightly smaller in size than benign ones (p = 0.005). Microcalcifications were present in 125 (50.8%) of the examined nodules, while the majority of them (113, 90.4%) were proven to be malignant (p < 0.001). Hypoechogenicity also showed a high correlation with malignancy ($p \leq 0.001$), as 89.3% of the hypoechoic nodules represented well-differentiated thyroid carcinomas (134 out of 150 hypoechoic nodules in total), while 61.0% of the malignant nodules were hypoechoic. Hyperechogenicity was rarely observed within both groups. Both irregular margins and taller-than-wide shapes were much more common in malignant nodules (92.2% vs. 7.8%, and 90.5% vs. 9.5%, respectively) and, thus, were highly associated with malignancy (p < 0.0001). In addition, almost 62% of the malignant nodules had a taller-than-wide shape (124 out of 201). No correlation was proven between malignancy and the vascularity pattern of the nodules in this study. All of the nodules that demonstrated extra-thyroidal extension on US evaluation proved to be thyroid carcinomas, but it was a rarely noted feature (4.9% of the nodules examined) (Table 2).

Table 2. Ultrasound characteristics of the thyroid nodules.

		All Nodules (*n* = 246)	WDTC Group (*n* = 201)	NC Group (*n* = 45)	*p*-Value
Median nodule size [mm] (IQR)		21.0 (17.3)	20.0 (10.0)	27.0 (60.0)	0.005
Microcalcifications (%)	Yes	125 (50.8)	113 (56.2)	12 (26.7)	<0.001
Hypervascularity of nodule (%)	Yes	155 (63.0)	122 (60.7)	33 (73.3)	0.112
Hyperechoic nodule (%)	Yes	10 (4.1)	3 (2)	7 (15.6)	<0.001
Hypoechoic nodule (%)	Yes	150 (61.0)	134 (66.7)	16 (35.6)	<0.001
Cystic elements (%)	Yes	72 (29.3)	50 (24.9)	22 (48.9)	0.001
Taller than wide (%)	Yes	137 (55.7)	124 (61.7)	13 (28.9)	<0.001
Extrathyroidal Extension (%)	Yes	12 (4.9)	12 (6.0)	0 (0.0)	0.093
Irregular margins (%)	Yes	103 (41.9)	95 (47.3)	8 (17.8)	<0.001

3.3. Multivariate Regression Analysis and Diagnostic Performances

Multivariate logistic regression analysis was performed to determine the malignancy prediction and the odds ratios for the US score (and its components separately) and the Bethesda System, respectively. In the final multivariate model, gender and age were also included for both models. Independent risk factors for malignancy were independently both US score (OR: 2.177; p < 0.0001) and Bethesda System (OR: 1.875; p = 0.002) (Tables 3 and 4). It is interesting to see that the characteristics included in the US score did not contribute equally. The risk score for irregular margins was more elevated than that of "taller-than-wide" or hypoechoic character of the nodule. On the other hand, the nodule size had risk score values near zero.

Table 3. Logistic models for the US score (WDTC group vs. NC group) prediction.

Parameter	Univariate Analysis					Multivariable Analysis					
	Unadjusted β	SE	OR	95% CI	*p*-Value	Adjusted β	SE	OR	95% CI	*p*-Value	RS
Gender (male/female)	0.257	0.405	1.29	0.59, 2.86	0.525						
Age at surgery (years)	−0.004	0.011	1.00	0.98, 1.02	0.746						
TSH pre-surgery	0.121	0.115	1.13	0.90, 1.41	0.293						
US score	0.788	0.157	2.200	1.616–2.994	<0.001	0.778	0.157	2.177	1.599–2.963	<0.001	
Nodule size	−0.026	0.011	0.97	0.95, 0.99	0.018	0.009	0.017	1.01	0.98, 1.04	0.606	0.01
Taller than wide	2.358	0.353	10.57	5.29, 21.11	<0.001	3.580	0.676	35.86	9.54, 134.83	<0.001	3.6
Irregular margins	2.703	0.372	14.92	7.20, 30.90	<0.001	3.925	0.678	50.67	13.41, 191.54	<0.001	3.9
Hypoechoic nodule	1.504	0.329	4.50	2.36, 8.57	<0.001	2.208	0.561	9.10	3.03, 27.31	<0.001	2.2

Abbreviations: β, coefficient of the explanatory variable; SE, standard error; OR, odds ratio; CI: confidence interval; RS: risk score.

Table 4. Logistic models for the Bethesda (WDTC group vs. NC group) prediction.

Parameter	Univariate Analysis					Multivariable Analysis					
	Unadjusted β	SE	OR	95% CI	*p*-Value	Adjusted β	SE	OR	95% CI	*p*-Value	RS
Gender (male/female)	−0.932	0.567	0.39	0.13, 1.20	0.100	−0.976	0.607	0.38	0.12, 1.24	0.108	−0.98
Age at surgery (years)	−0.029	0.013	0.97	0.95, 0.99	0.021	−0.021	0.014	0.98	0.95, 1.01	0.124	−0.02
TSH pre-surgery	0.106	0.130	1.11	0.86, 1.43	0.415						
Bethesda System	0.677	0.207	1.969	1.312–2.951	0.001	0.629	0.484	1.875	1.257–2.798	0.002	
Nodule size	−0.052	0.015	0.95	0.92, 0.98	<0.001	−0.039	0.016	0.96	0.93, 0.99	0.013	−0.04
Taller than wide	−0.029	0.337	0.97	0.50, 1.88	0.931						
Irregular margins	0.782	0.336	2.19	1.13, 4.22	0.020	0.551	0.364	1.74	0.85, 3.54	0.131	0.55
Hypoechoic nodule	0.586	0.367	1.80	0.87, 3.69	0.111	0.406	0.415	1.50	0.67, 3.39	0.328	0.41

Abbreviations: β, coefficient of the explanatory variable; SE, standard error; OR, odds ratio; CI: confidence interval; RS: risk score.

The discriminatory performance of US score for predicting thyroid nodule malignancy and the respective ROC curve is presented in Figure 1. More specifically, the US score exhibited a significant strong discriminatory performance (AUC = 0.784, CI 95%: 0.693–0.875). Regarding the optimal cut-off point of the US score, this was 3.5, with sensitivity of 79.9% and specificity of 66.7%.

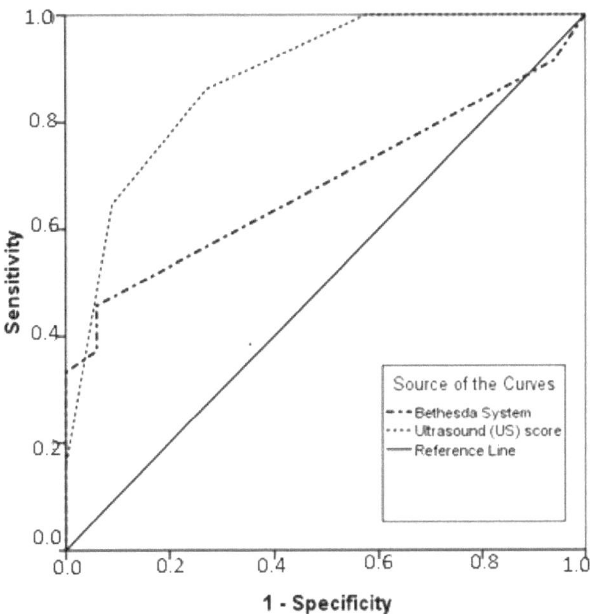

Figure 1. Receiver operating characteristic (ROC) curves of the Bethesda System and US score.

3.4. Diagnostic Performances

Diagnostic tests were performed to evaluate the sensitivity, specificity, positive predictive value and negative predictive value of both the US scoring system and the Bethesda System in this study.

The two methods were compared in regard to their diagnostic accuracy, sensitivity, specificity and positive and negative predictive value. While the Bethesda System demonstrated specificity and PPV of 100%, its sensitivity was proven to be as low as 33%. On the other hand, the US scoring system was found to have a lower specificity rate and PPV (90.9% and 97.1%, respectively), but with a much higher sensitivity value (64.2%) and a higher NPV (34.5% vs. 24.4%) with a statistical significance of $p < 0.001$ (Table 5). The diagnostic accuracy was 68.75% for the US scoring system and 43.75% for the Bethesda System. A ROC curve analysis was performed for the evaluation of both systems, and the AUC was calculated. The reference standard for malignancy was the histopathology report. When comparing the two AUCs with a Z score test, a statistically significant difference between the two methods was observed. The AUC for the US score ROC curve analysis was 88.0%, a significantly higher value compared to the AUC of the Bethesda System ROC curve analysis with a value of 68.3% ($p < 0.001$) (Figure 1).

Table 5. Diagnostic accuracy of US score and Bethesda System in the diagnosis of cancer.

		Group WDTC	Group NC	Sensitivity	Specificity	PPV	NPV
US score	Positive [3–5]	102	3	64.2%	90.9%	97.1%	34.5%
	Negative [0–2]	57	30				
Bethesda	Positive [5–6]	51	0	33.3%	100%	100%	24.4%
	Negative [1–4]	102	33				
p-value				<0.001	0.083	0.083	<0.001

Abbreviations: PPV, positive predictive value; NPV, negative predictive value.

4. Discussion

The diagnostic evaluation of thyroid nodules is a difficult process. It involves a careful history and clinical examination, followed by a thyroid ultrasound and hormonal tests to assess the thyroid function and the presence of autoantibodies. The clinical importance of thyroid nodules hinges on the need to diagnose thyroid cancer, which occurs in 7–15% of cases based on age, sex, radiation exposure history, family history, smoking habit, obesity and other factors [5,11]. Nowadays, about 40% of the WDTCs diagnosed are less than 1 cm. This tumor shift may be due to the increasing use of ultrasonography or other imaging methods and early diagnosis and treatment [12,13]. In a large retrospective study by Chen et al., the higher incidence of thyroid cancer in thyroid nodules screened with ultrasound rather than palpation was established, and thus, two-thirds of the thyroid nodules believed to be normal were microcarcinomas [13]. The optimization of long-term health outcomes and education about potential prognoses for individuals with thyroid neoplasms is critically important.

In the international literature, there are some systematic reviews and meta-analyses that tried to analyze ultrasound and FNA diagnostic accuracy in thyroid malignancy [14,15]. Ospina et al. concluded that the available evidence only warrants limited confidence on the diagnostic accuracy of FNA due to risk of bias, imprecision and inconsistency among studies, but the likelihood for purely benign and purely malignant potential was found to be high [15]. Regarding ultrasound, Remonti et al. reported that solitary ultrasonographic findings alone could not predict malignancy, but the combination of microcalcifications, a taller-than-wide shape, irregular margins or the absence of elasticity could offer reliable information in terms of malignant potential [14]. The absence of elasticity had the best ultra-sonographic performance for malignant results [14]. The study by Ito et al. demonstrated a positive predictive value of ultrasound in 97.2%, while Nie et al. reported ultrasound as being a highly accurate examination for thyroid nodule nature discrimination with a specificity of 33.88% and sensitivity of 92.53% [16,17]. In terms of cytology, FNA was found to have positive predictive value of 100% and negative predictive value of 43.75% in a randomized cross-sectional study [18]. Thus, our study tried to interpret the diagnostic accuracy of ultrasound and FNA in thyroid malignancy diagnosis retrospectively. Our results demonstrated ultrasonographic findings such as microcalcifications, hypoechogenicity, irregular margins and taller-than-wide shapes to be correlated to malignancies with high statistical significance. Furthermore, our study is in concordance with the international available data in terms of positive and negative predictive values of FNA and ultrasound scores, while we concluded that ultrasound specificity was higher and sensitivity was lower than the values proposed in other studies.

Several studies have reported the utility of ultrasonography alone for distinguishing benign from malignant nodules [6,19]. In addition, it is cost-effective, widely available and not invasive. Therefore, ultrasonography has been adopted as the first useful step in determining the location and nature of thyroid nodules [16,20–24]. Alshoabi et al. reported that B-mode ultrasonography alone could differentiate benign nodules with excellent diagnostic accuracy [20]. In this context, we tried to separately examine the sonographic characteristics of thyroid nodules. Their characteristics together with the total US score were analyzed. Moreover, we analyzed the Bethesda score and compared the diagnostic accuracy of the scores. From the uni- and multivariate analysis of both the Bethesda and US scoring systems, we can safely conclude that both could effectively predict the existence of a TN containing a malignancy. Interestingly, the US score displayed higher sensitivity and better negative predictive value than the Bethesda score, while both scoring systems displayed comparable specificities and positive predictive values. Not surprisingly, in the present study, the diagnostic accuracy of the US score was superior to Bethesda score.

When looking closely at the US score and its' various components, we observed that microcalcification, hypoechogenicity, cystic element, "taller-than-wide" and irregular margins are parameters that are separately strongly correlated with the malignant potential of the TN. This is in consistency with other studies in the literature. Nabahati et al. and

Ram et al. proposed the same parameters as indicators of malignancy [19,21]. This means—on a clinical basis—that a patient presenting with one of those characteristics should be considered as having a malignancy and treated adequately.

Moreover, the uni- and multivariate analysis of the logistic regression model of US scoring showed that regardless of the other characteristics, irregular margin, hypoechoic nodule and "taller-than-wide" are strong predictors of malignancy. On the contrary, the nodule size seems not to have a role in the malignant potential of a TN, which is in agreement with Rahimi et al.'s study, concluding that nodule size should not be a criterion for malignancy, but irregular edges, being solid, hypoechogenicity and being a single nodule are major components of malignancy [24]. In the analysis of the Bethesda System, nodule size seems to have a pivotal predictive role in malignancy. Thyroid cytology faces pitfalls in false positive and negative results, such as the misinterpretation of cystic degeneration and squamous cells in partially cystic lesions or the underestimation of architectural and cellular features in follicular patterns, with a wide range of sensitivity (65–99%) and specificity (72–100%) [25]. Zhu et al. resulted that sampling error (86.7%) was the most common cause of false negative diagnoses in FNA, mainly due to nodule size, while interpretation error (80.9%) was the most common cause of false positive diagnoses, affected by overlapping cytological features in adenomatous hyperplasia, thyroiditis and cystic lesions [26]. In our study, the Bethesda System demonstrated specificity and PPV of 100%, while its sensitivity was proven to be as low as 33% and its diagnostic accuracy was 43.75%.

This study has critical clinical implications. First of all, the results of our retrospective study reinforce the fact that ultrasound and FNA biopsy could lead the decision behind the surgical management of suspicious thyroid nodules. Ultrasound has high specificity and positive predictive value for malignant potential, while we found that ultrasound sensitivity is less than already proposed values. FNA biopsy also features high specificity and positive predictive value for thyroid malignancy. Compared to one another, especially in terms of sensitivity and negative predictive values which are lacking in both examinations, ultrasound was found to score higher than FNA biopsy. This is a pivotal new finding that could provide advanced perspectives in the management of suspicious thyroid nodules. In terms of indetermined results between ultrasound and FNA biopsy, we propose that ultrasound findings are more reliable than FNA biopsy of suspicious thyroid nodules, based on the findings of a fully experienced radiologist in the field of the thyroid gland.

5. Limitations

A major limitation of our study is the retrospective form of collection of data. We had strict inclusion criteria to lower the risk of selection bias.

6. Conclusions

The diagnostic evaluation of thyroid nodules is a difficult process. Sonographic findings dictate fine-needle aspiration cytology and combined approaches lead the diagnostic process of thyroid malignancy. Our study demonstrates that US score has higher sensitivity and better negative predictive value than Bethesda score, while both scoring systems displayed comparable specificities and positive predictive values. Ultrasound seems to be more reliable in predicting thyroid malignancy, with microcalcifications, hypoechogenicity, irregular margins, and taller-than-wide shapes being major factors of malignant potential.

Author Contributions: Study conception and design: M.B., T.P. and M.K.; data collection: E.T., A.C. (Alexandra Chrisoulidou), A.C. (Angeliki Chorti), G.T., R.V. and P.P.; analysis and interpretation of results: M.B., A.C. (Alexandra Chrisoulidou) and E.T.; draft manuscript preparation: A.C. (Angeliki Chorti), M.B., A.C. (Alexandra Chrisoulidou), A.I. and E.T.; revision of the manuscript: M.B., T.P. and M.K. All authors have read and agreed to the published version of the manuscript.

Funding: This review has not been supported by any grants or funds.

Institutional Review Board Statement: The study was conducted in accordance with the Declaration of Helsinki and approved by the Institutional Review Board (or Ethics Committee) of Theageneio Cancer Hospital (2661/22 February 2024).

Informed Consent Statement: Informed consent was obtained from all subjects involved in the study.

Data Availability Statement: Data will be available on reasonable request.

Conflicts of Interest: The authors declare no conflict of interest.

References

1. Dean, D.S.; Gharib, H. Epidemiology of thyroid nodules. *Best. Pract. Res. Clin. Endocrinol. Metab.* **2008**, *22*, 901–911. [CrossRef]
2. Rossi, E.D.; Pantanowitz, L.; Hornick, J.L. A worldwide journey of thyroid cancer incidence centred on tumour histology. *Lancet Diabetes Endocrinol.* **2021**, *9*, 193–194. [CrossRef]
3. Chieng, J.; Lee, C.; Karandikar, A.; Goh, J.; Tan, S. Accuracy of ultrasonography-guided fine needle aspiration cytology and significance of non-diagnostic cytology in the preoperative detection of thyroid malignancy. *Singap. Med. J.* **2019**, *60*, 193–198. [CrossRef]
4. Ali, S.Z.; Baloch, Z.W.; Cochand-Priollet, B.; Schmitt, F.C.; Vielh, P.; VanderLaan, P.A. The 2023 Bethesda System for Reporting Thyroid Cytopathology. *Thyroid* **2023**, *33*, 9. [CrossRef]
5. Haugen, B.R.; Alexander, E.K.; Bible, K.C.; Doherty, G.M.; Mandel, S.J.; Nikiforov, Y.E.; Pacini, F.; Randolph, G.W.; Sawka, A.M.; Schlumberge, M.; et al. 2015 American Thyroid Association Management Guidelines for Adult Patients with Thyroid Nodules and Differentiated Thyroid Cancer: The American Thyroid Association Guidelines Task Force on Thyroid Nodules and Differentiated Thyroid Cancer. *Thyroid* **2016**, *26*, 1–133. [CrossRef]
6. Lee, Y.H.; Baek, J.H.; Jung, S.L.; Kwak, J.Y.; Kim, J.; Shin, J.H. Ultrasound-Guided Fine Needle Aspiration of Thyroid Nodules: A Consensus Statement by the Korean Society of Thyroid Radiology. *Korean J. Radiol.* **2015**, *16*, 391. [CrossRef]
7. Russ, G.; Bonnema, S.J.; Erdogan, M.F.; Durante, C.; Ngu, R.; Leenhardt, L. European Thyroid Association Guidelines for Ultrasound Malignancy Risk Stratification of Thyroid Nodules in Adults: The EU-TIRADS. *Eur. Thyroid. J.* **2017**, *6*, 225–237. [CrossRef]
8. Weller, A.; Sharif, B.; Qarib, M.H.; St Leger, D.; De Silva, H.S.; Lingam, R.K. British Thyroid Association 2014 classification ultrasound scoring of thyroid nodules in predicting malignancy: Diagnostic performance and inter-observer agreement. *Ultrasound* **2020**, *28*, 4–13. [CrossRef]
9. Perkins, N.J.; Schisterman, E.F. The Inconsistency of "Optimal" Cutpoints Obtained using Two Criteria based on the Receiver Operating Characteristic Curve. *Am. J. Epidemiol.* **2006**, *163*, 670–675. [CrossRef] [PubMed]
10. Youden, W.J. Index for rating diagnostic tests. *Cancer* **1950**, *3*, 32–35. [CrossRef] [PubMed]
11. Hegedüs, L. The Thyroid Nodule. *N. Engl. J. Med.* **2004**, *351*, 1764–1771. [CrossRef]
12. Leenhardt, L.; Bernier, M.; Boin-Pineau, M.; Conte Devolx, B.; Marechaud, R.; Niccoli-Sire, P.; Nocaudie, M.; Orgiazzi, J.; Schlumberger, M.; Wemeau, J.L.; et al. Advances in diagnostic practices affect thyroid cancer incidence in France. *Eur. J. Endocrinol.* **2004**, *150*, 133–139. [CrossRef] [PubMed]
13. Chen, Z.; Mosha, S.S.; Zhang, T.; Xu, M.; Li, Y.; Hu, Z.; Liang, W.; Deng, X.; Ou, T.; Li, L.; et al. Incidence of microcarcinoma and non-microcarcinoma in ultrasound-found thyroid nodules. *BMC Endocr. Disord.* **2021**, *21*, 38. [CrossRef]
14. Remonti, L.R.; Kramer, C.K.; Leitão, C.B.; Pinto, L.C.F.; Gross, J.L. Thyroid Ultrasound Features and Risk of Carcinoma: A Systematic Review and Meta-Analysis of Observational Studies. *Thyroid* **2015**, *25*, 538–550. [CrossRef]
15. Singh Ospina, N.; Brito, J.P.; Maraka, S.; Espinosa de Ycaza, A.E.; Rodriguez-Gutierrez, R.; Gionfriddo, M.R.; Castaneda-Guardares, A.; Benkhadra, K.; Al Nofal, A.; Erwin, P.; et al. Diagnostic accuracy of ultrasound-guided fine needle aspiration biopsy for thyroid malignancy: Systematic review and meta-analysis. *Endocrine* **2016**, *53*, 651–661. [CrossRef]
16. Ito, Y.; Amino, N.; Yokozawa, T.; Ota, H.; Ohshita, M.; Murata, N.; Morita, S.; Kobayashi, K.; Miyaguci, A. Ultrasonographic Evaluation of Thyroid Nodules in 900 Patients: Comparison Among Ultrasonographic, Cytological, and Histological Findings. *Thyroid* **2007**, *17*, 1269–1276. [CrossRef]
17. Nie, W.; Zhu, L.; Yan, P.; Sun, J. Thyroid nodule ultrasound accuracy in predicting thyroid malignancy based on TIRADS system. *Adv. Clin. Exp. Med.* **2022**, *31*, 597–606. [CrossRef] [PubMed]
18. Servilla, M.; Trabanco, C.; Feliciano, W.; Bredy, R.E.; Lojo, S. Association between fine needle aspiration cytology and final pathology in the diagnosis of thyroid nodules with surgical indications. *Bol. Asoc. Med. P. R.* **2016**, *108*, 13–18.
19. Ram, N.; Hafeez, S.; Qamar, S.; Hussain, S.Z.; Asghar, A.; Anwar, Z.; Islam, N. Diagnostic validity of ultrasonography in thyroid nodules. *J. Pak. Med. Assoc.* **2015**, *65*, 875–878. [CrossRef]
20. Alshoabi, S.A.; Binnuhaid, A.A. Diagnostic accuracy of ultrasonography versus fine-needle-aspiration cytology for predicting benign thyroid lesions. *Pak. J. Med. Sci.* **2019**, *35*, 630. [CrossRef]
21. Nabahati, M.; Moazezi, Z.; Fartookzadeh, S.; Mehraeen, R.; Ghaemian, N.; Sharbatdaran, M. The comparison of accuracy of ultrasonographic features versus ultrasound-guided fine-needle aspiration cytology in diagnosis of malignant thyroid nodules. *J. Ultrasound* **2019**, *22*, 315–321. [CrossRef] [PubMed]

22. Xie, C.; Cox, P.; Taylor, N.; LaPorte, S. Ultrasonography of thyroid nodules: A pictorial review. *Insights Imaging* **2016**, *7*, 77–86. [CrossRef] [PubMed]
23. Frates, M.C.; Benson, C.B.; Doubilet, P.M.; Kunreuther, E.; Contreras, M.; Cibas, E.S.; Orcutt, J.; Moore, F.D.; Larsen, P.R.; Marquesse, E.; et al. Prevalence and Distribution of Carcinoma in Patients with Solitary and Multiple Thyroid Nodules on Sonography. *J. Clin. Endocrinol. Metab.* **2006**, *91*, 3411–3417. [CrossRef] [PubMed]
24. Rahimi, M.; Farshchian, N.; Rezaee, E.; Shahebrahimi, K.; Madani, H. To differentiate benign from malignant thyroid nodule comparison of sonography with FNAC findings. *Pak. J. Med. Sci.* **2012**, *29*, 77. [CrossRef]
25. Rossi, E.D.; Adeniran, A.J.; Faquin, W.C. Pitfalls in Thyroid Cytopathology. *Surg. Pathol. Clin.* **2019**, *12*, 865–881. [CrossRef]
26. Zhu, Y.; Song, Y.; Xu, G.; Fan, Z.; Ren, W. Causes of misdiagnoses by thyroid fine-needle aspiration cytology (FNAC): Our experience and a systematic review. *Diagn. Pathol.* **2020**, *15*, 1. [CrossRef]

 *medicina*

Article

Frequency of Thyroid Microcarcinoma in Patients Who Underwent Total Thyroidectomy with Benign Indication—A 5-Year Retrospective Review

Vasiliki Magra [1,*], Kassiani Boulogeorgou [2], Eleni Paschou [1], Christina Sevva [1], Vasiliki Manaki [3], Ioanna Mpotani [3], Stylianos Mantalovas [1], Styliani Laskou [1], Isaak Kesisoglou [1], Triantafyllia Koletsa [2] and Konstantinos Sapalidis [1]

[1] Third Department of General Surgery, Medical School, AHEPA University Hospital, Aristotle University of Thessaloniki, 54621 Thessaloniki, Greece; paschoueleni@gmail.com (E.P.); christina.sevva@gmail.com (C.S.); steliosmantalobas@yahoo.gr (S.M.); ikesis@hotmail.com (I.K.); sapalidiskonstantinos@gmail.com (K.S.)

[2] Department of Pathology, Faculty of Medicine, School of Health Sciences, University Campus, Aristotle University of Thessaloniki, 54124 Thessaloniki, Greece; siliaboulog@gmail.com (K.B.); tkoletsa@auth.gr (T.K.)

[3] School of Medicine, Aristotle University of Thessaloniki, 54124 Thessaloniki, Greece; vassiamanaki@gmail.com (V.M.); ioannampo@hotmail.co.uk (I.M.)

* Correspondence: valia.magra@gmail.com

Abstract: *Background and Objectives*: Incidental thyroid cancers (ITCs) are often microcarcinomas. The most frequent histologic type is a papillary microcarcinoma. Papillary thyroid microcarcinomas are defined as papillary thyroid tumours measuring less than 10 mm at their greatest diameter. They are clinically occult and frequently diagnosed incidentally in histopathology reports after a thyroidectomy. The aim of this study is to evaluate the rate of papillary thyroid microcarcinomas (PTMC) in patients who were thyroidectomised with indications of benign disease. *Materials and Methods*: We retrospectively evaluated the histological incidence of PTMC in 431 consecutive patients who, in a 5 year period, underwent a thyroidectomy with benign indications. Patients with benign histology and with known or suspected malignancy were excluded. *Results*: Histopathology reports from 540 patients who underwent a total thyroidectomy in our department between 2016 and 2021 were reviewed. A total of 431 patients were thyroidectomised for presumed benign thyroid disease. A total of 395 patients had confirmed benign thyroid disease in the final histopathology, while 36 patients had incidental malignant lesions (33 PTMC—7.67%, one multifocal PTC without microcarcinomas—0.23%, two follicular thyroid carcinoma—0.46%). Out of the PTMC patients, 29 were female and four were male (7.2:1 female–male ratio). The mean age was 54.2 years old. A total of 24 out of 33 patients had multifocal lesions, 11 of which co-existed with macro PTC. Nine patients had unifocal lesions. A total of 21 of these patients were initially operated on for multinodular goitre (64%), while 13 were operated on for Hashimoto/Lymphocytic thyroiditis (36%). *Conclusions*: PTMC—often multifocal—is not an uncommon, incidental finding after thyroidectomy for benign thyroid lesions (7.67% in our series) and often co-exists with other incidental malignant lesions (8.35% in our series). The possibility of an underlying papillary microcarcinoma should be taken into account in the management of patients with benign—especially nodular—thyroid disease, and total thyroidectomy should be considered.

Keywords: thyroidectomy; papillary; microcarcinoma; incidental; thyroid

Citation: Magra, V.; Boulogeorgou, K.; Paschou, E.; Sevva, C.; Manaki, V.; Mpotani, I.; Mantalovas, S.; Laskou, S.; Kesisoglou, I.; Koletsa, T.; et al. Frequency of Thyroid Microcarcinoma in Patients Who Underwent Total Thyroidectomy with Benign Indication—A 5-Year Retrospective Review. *Medicina* **2024**, *60*, 468. https://doi.org/10.3390/medicina60030468

Academic Editor: Fumio Otsuka

Received: 10 February 2024
Revised: 27 February 2024
Accepted: 11 March 2024
Published: 12 March 2024

1. Introduction

Thyroid cancer is the most frequent type of endocrine cancer. A papillary thyroid carcinoma is the most frequently encountered thyroid tumour and comprises 70–90% of well-differentiated thyroid cancers [1]. Incidentally found thyroid carcinomas imply lesions that were not clinically suspected prior to their diagnosis. In previous years, these carcinomas were all considered incidentalomas and were not specifically categorised. More recently, they were thought of as a separate entity and were classified into four broad categories based on the way they are diagnosed [2]: (i) microcarcinomas diagnosed incidentally in histopathology reports post thyroidectomy despite no clinical suspicion for malignancy pre-operatively, (ii) microcarcinomas incidentally detected during imaging, usually ultrasound, which were confirmed with fine needle aspiration cytology (FNAC), (iii) carcinomas that first present with lymph node metastases and the primary lesion—a microcarcinoma—is identified in post-operative histopathology, and (iv) carcinomas in ectopic thyroid tissue that present with metastases or clinical symptomatology [2]. In clinical practice, these microcarcinomas are most frequently found in histopathology reports of thyroidectomy specimens in patients with benign indications for surgery [3]. The most frequent histotype encountered is a papillary thyroid microcarcinoma (PTMC).

The World Health Organisation (WHO) defines thyroid microcarcinoma as thyroid carcinoma with a maximum diameter under 1 cm and without including any further tumour characteristics, such as the presence of metastatic lesions or local extra-thyroidal extension of the tumour [4]. The recently published WHO classification of thyroid tumours (2022) notes that a papillary thyroid carcinoma should be classified based on the tumour's histopathologic characteristics, irrespective of size. It is also stressed that tumours under 1 cm in diameter should not be considered low-risk simply because of their size [5]. Papillary thyroid microcarcinomas are usually multifocal and most frequently found in thyroid glands with a multinodular pathology.

The improved sensitivity of imaging modalities, especially ultrasound, as well as the increased accuracy of histopathology examination of the thyroid specimen, can explain the increased incidence of incidentally found thyroid microcarcinomas [1]. The incidence of papillary thyroid microcarcinomas in autopsy and clinical studies is 100- to 1000-fold higher than the incidence of such tumours that develop into clinically apparent cancers encountered in clinical practice [2]. It is, therefore, reasonable to consider if the diagnosis of a papillary microcarcinoma should automatically become an indication for surgical resection.

The clinical significance of papillary thyroid microcarcinomas and their clinical course remains a matter of interest and debate. Papillary thyroid microcarcinomas do not appear to be a unified entity. The clinical progress of these lesions is not yet clear or specific. While the majority have an excellent prognosis, with a 20 year mortality rate of less than 1%, they do have metastatic potential to neck lymph nodes (via lymphatic spread) and less frequently to distant sites (via haematogenous spread). It has also been observed that 4–16% of papillary thyroid microcarcinomas recur locally after initial excision, and a significant percentage of these patients develop distant metastases [6]. Roti et al. tried to identify the clinical characteristics that differentiated incidentally diagnosed papillary microcarcinomas from thyroid lesions suspicious of malignancy. The only differentiating factor between the two groups was size. No lesions smaller than 8 mm in diameter developed metastases [7].

At this point, it is important to mention a subtype of papillary thyroid microcarcinoma, an aggressive PTMC. Ardito et al. conducted a retrospective study on papillary thyroid microcarcinoma and concluded that although most of them had uncomplicated clinical progress, 19% recurred [8]. The microcarcinomas that recurred had common histopathologic characteristics. More specifically, they were multifocal, lacked a capsule, and were solid. As a result, they suggested that this subtype of papillary thyroid microcarcinomas should be treated aggressively with a total thyroidectomy, central lymph node clearance, and radioiodine [8].

There is currently no consensus regarding the clinical progression of papillary thyroid microcarcinomas or the most appropriate treatment strategy. Options range from watch and wait to a total thyroidectomy with central lymph node clearance and radioiodine, as in the patients from the aforementioned study. The American Thyroid Association 2015 guidelines suggest that for cN0 tumours < 1 cm without extrathyroidal extensions, a hemithyroidectomy should be the operation of choice [9]. Despite this guideline, many authors suggest that a total thyroidectomy and central lymph node clearance should be the treatment strategy of choice due to the frequency of multifocality of such tumours as well as their tendency to metastasise to level IV lymph nodes in the neck. This avoids re-operation and reduces the risk of local recurrence. Spaziani et al. [10] published a recent prospective observational study that identified a high percentage of incidental papillary thyroid microcarcinomas (15.3%) with high rates of bilateral lesions and multifocality (63.1%). They go on to suggest a total thyroidectomy as the surgical treatment of choice, even for presumed benign thyroid disease [10]. The main counterargument is the increased frequency of central lymph node clearance, especially permanent hypocalcemia [11].

This retrospective study aims to investigate the frequency and multifocality of incidentally diagnosed thyroid papillary microcarcinomas within the thyroid gland. The literature suggests that the incidence of such tumours ranges from 3% to 16%, although, occasionally, much higher numbers have been documented (up to 81%) [1]. Keeping in mind the high incidence of these incidental microcarcinomas, as well as their tendency for multifocality, a total thyroidectomy seems to be a reasonable treatment strategy, especially in cases of underlying multinodular goitre when both thyroid lobes are affected.

All the patients included in this study underwent a total thyroidectomy. Given that the main counterargument against a total thyroidectomy vs a hemithyroidectomy for benign thyroid lesions is the increased rate of complications, we mention the postoperative complications in the patient cohort studied and proceed to compare them to the international literature.

2. Materials and Methods

This is a 5 year retrospective study for which we retrieved 540 histopathology reports of patients that underwent a total thyroidectomy at the 3rd General Surgical Department of AHEPA University Hospital from 2016 until 2021. A total of 145 histopathology reports confirmed thyroid malignancies, while 395 reports confirmed benign thyroid lesions (multinodular goitre, Graves disease, and Hashimoto thyroiditis). A total of 58 papillary microcarcinomas were confirmed, 33 of which were incidental findings in patients who underwent a total thyroidectomy with benign pre-operative indication. These were divided into unifocal and multifocal; their mean diameter and the standard deviation (SD) were calculated, as well as the number of lesions in multifocal carcinomas.

Patients with a confirmed pre-operative diagnosis of malignancy (post-fine needle aspiration cytology) were excluded from the study. Patients with histopathology that confirmed exclusively benign thyroid diagnoses were also excluded from the study. Finally, 2 patients with incidentally diagnosed follicular thyroid carcinomas were excluded. All patients underwent a preoperative endocrinology assessment, clinical examination of the neck, laboratory examination for thyroid hormones (TSH, T3, T4, procalcitonin, anti-thyroid peroxidase (anti-TPO), and thyroglobulin), neck/thyroid ultrasound, pre-operative laryngoscopy, and, based on indication, a neck computed tomography (CT) scan and fine needle aspiration (FNA). Intraoperative nerve monitoring was used in all operations.

There was telephone contact with 431 patients who underwent a total thyroidectomy due to presumed benign thyroid disease to confirm complications and recurrence. The histopathology reports of all patients were also reviewed, and parathyroid glands were included. Complications were divided into immediate/transient and permanent.

3. Results

From a total of 540 patients, 431 had a preoperative benign indication for total thyroidectomy (395 had confirmed benign diagnosis in the histopathology report of the thyroidectomy specimen, 33 had incidentally diagnosed papillary microcarcinomas, one had incidentally diagnosed papillary macro carcinoma and two had incidentally diagnosed follicular carcinomas). In total, 33 out of 431 (7.67%) patients were found to have papillary thyroid microcarcinomas. These patients had benign indications for thyroidectomy pre-operatively and no suspicion of thyroid malignancy during the pre-operative work-up. These patients constitute the focus of this study.

Out of the 33 patients with incidentally diagnosed papillary microcarcinomas, the pre-operative diagnosis for 20 of them (60.6%) was a multinodular goitre, and for the remaining 13 patients (39.4%), Hashimoto thyroiditis. These diagnoses were confirmed in the histopathology reports post-operatively. There were no papillary microcarcinomas in the histopathology specimen of patients with Graves disease. A total of 29 patients were women, and four were men (7.2:1 F–M ratio). The mean age for women was 54.2 years old (31–72) and 54.2 years old (50–59) for men.

A total of 24 patients (72.7%) had multifocal papillary microcarcinomas (23 women with a mean age of 52.2 years old and one 52 year old man). Nine patients (27.3%) had unifocal papillary microcarcinomas (six women with a mean age of 61.8 years old and three men with a mean age of 55 years old). The mean maximal diameter of multifocal microcarcinomas was 3.5 mm $\pm$ 0.19 mm SD, and for unifocal microcarcinomas was 3.5 mm $\pm$ 0.23 mm SD. The mean number of lesions in multifocal microcarcinomas was 2.83 (2–6 lesions per patient). In 11 (45.8%) out of 24 patients, multifocal papillary microcarcinomas co-existed with papillary thyroid carcinomas (macro PTC > 1 cm).

There was telephone communication with 431 patients who underwent a total thyroidectomy with a benign indication for surgery. Contact was not achieved with eight of these patients. Patients were asked about immediate/transient and permanent post-operative complications, specifically about hypoparathyroidism and vocal cord palsy symptoms (numbness, hoarseness of voice, stridor, breathing difficulty, and emergency tracheostomy). A total of 11 out of 423 patients (2.60%) reported transient hypocalcaemia, three reported permanent hypocalcaemia (0.70%), and four reported transient unilateral vocal cord paresis (0.94%), which normalised in repeat laryngoscopy one year post-operatively. No patients reported stridor or required an emergency tracheostomy due to bilateral vocal cord paresis. Parathyroid glands were found in 5 out of 431 (1.16%) histopathology reports, four of which contained one parathyroid gland, while 1 contained two parathyroid glands. There were no cases of local recurrence, lymphatic spread, or distant metastases to date. Table 1 below shows the characteristics of the patients and the microcarcinomas, focusing on the multifocality of the tumour.

Table 1. Characteristics of patients with papillary thyroid microcarcinoma.

Characteristics	Female	Male	Total
Multifocal tumours	23	1	24
Unifocal tumours	6	3	9
Mean age	54.2 years old (31–72)	54.2 years old (50–59)	n/a
Multinodular goitre	16	4	20
Hashimoto thyroiditis	13	0	13

4. Discussion

Papillary thyroid microcarcinomas are a unique subdivision of papillary thyroid carcinomas, which heralds attention due to their increasing frequency as well as the implications for patient management [12]. Although currently considered a unified category of thyroid tumours, many authors have started to consider papillary thyroid microcarcinomas as a group of tumours that may exhibit different biological behaviours. This would explain why some microcarcinomas have an indolent course while others behave aggressively. Different studies have suggested many risk factors for the more aggressive subdivision of papillary thyroid microcarcinomas without a unified opinion in the literature.

Papillary thyroid microcarcinomas and truly incidental thyroid carcinomas do not always belong to the same category. It is possible to detect incidental microcarcinomas in patients with an already-known thyroid carcinoma. This group of microcarcinomas would not be strictly defined as incidental. In this study, 63.6% of patients with incidental thyroid papillary microcarcinomas had a preoperative diagnosis of multinodular goitre. This finding is in accordance with reports in the literature, as microcarcinomas are encountered more frequently in nodular thyroid glands. The incidental diagnosis of a PTMC in patients with multinodular goitre requires frequent monitoring. The management of such patients should be dictated by the characteristics of the lesion as well as the age and sex of the patient, which Bove et al. found to be the most decisive factor [13]. These patients should be regularly monitored due to their high risk of developing papillary thyroid carcinomas.

Multiple authors have investigated the link between Hashimoto thyroiditis and an increased risk of developing thyroid cancer, especially papillary thyroid cancer. In this study, 39.4% of the patients incidentally diagnosed with a PTMC had histopathologically confirmed Hashimoto thyroiditis (HT)/chronic lymphocytic thyroiditis (CLT). Hu et al. [14] performed a systematic review and meta-analysis of the cancer risk in Hashimoto's thyroiditis and found an increased risk of developing thyroid cancer (RR = 1.49 $p < 0.067$). The increased risk is attributed to chronic inflammation [14–18], especially through the release of free oxygen radicals, inflammatory cytokines [14], and the expression of oncogenes [15,18]. Unger et al. [15] found a potentially linked expression of p63 in patients with Hashimoto's thyroiditis and patients with papillary thyroid carcinomas. Jackson et al. [16] also found a 1.5-fold increased risk of developing PTCs in patients with chronic lymphocytic thyroiditis while conducting a retrospective study of 1268 patients. Vargas-Uricoechea [17] notes a potential pathobiological link between autoimmune thyroid disease (Graves and HT) and thyroid cancer, especially papillary thyroid carcinoma, during their review of the literature. They stress, however, that the existing evidence from observational studies is not enough to establish causality [17]. Interestingly, they mention that the category of patients with "indeterminate" nodules (Bethesda III and IV) may benefit from better classification by using molecular markers [17]. Finally, Lee et al. [18] conducted a 9 year retrospective study of 2928 patients and found that chronic lymphocytic thyroiditis correlated positively with multifocality, smaller tumour size, extrathyroidal extension, and p53 expression but negatively with lymph node metastases.

Due to the uncertain clinical progression of papillary thyroid microcarcinomas, communication with the patient is key so that they are fully informed about the risks and benefits of different management options—surgical vs. active monitoring—and that they can be actively involved in the decision-making process [19]. Fine needle aspiration is known to be useful in the diagnosis of thyroid nodules. However, the role of fine needle aspiration cytology (FNAC) in patients with multinodular goitre remains controversial, especially when aiming to diagnose microcarcinomas. Lasithiotakis et al. reported suspicious FNA findings for carcinomas in 6.7% of cases [19], while Baier et al. had a 12% rate of false negative samples [20]. Miccoli et al. found the role of fine needle aspiration cytology in their study of incidental thyroid carcinomas to be inconclusive, even for tumours larger than 2 cm [21]. Evranos et al. concluded that cytologically suspicious nodules proved to be benign in surgical pathology, while cancer foci were identified in previously undefined areas of the thyroid in patients with incidentally diagnosed carcinomas [22]. None of the 33 patients that were included in this study had suspicious features on imaging and thus did not undergo FNA.

The size of microcarcinomas poses challenges in diagnosis both when imaging and when performing fine needle aspiration on such lesions. In the current study, the mean maximal diameter of multifocal PTMC and unifocal PTMC was 3.5 mm $\pm$ 0.19 SD and 3.5 mm $\pm$ 0.23 SD, respectively. A total of 11 patients had co-existing papillary thyroid carcinomas > 1 cm in diameter. It is evident that in cases of underlying multinodular goitre, the challenges are increased due to the morphology of the underlying gland. Lucandri et al. support this as they found that the mean diameter of non-incidental papillary thyroid microcarcinomas is double when compared to the incidental PTMC group (7.2 vs. 3.8 mm; $p < 0.001$) [23].

In most cases of patients with papillary thyroid microcarcinoma, the life expectancy is excellent, with a mortality of <1% in 20 years [1]. Despite the excellent life expectancy, papillary thyroid microcarcinomas often metastasise to cervical lymph nodes and occasionally to distant sites [1]. In the 4–16% of patients that have a local recurrence, the risk for distant metastasis is significantly higher [1]. Kaliszewski et al. [24] investigated the high-risk characteristics of papillary thyroid microcarcinomas in order to aid in guiding the management of such patients, especially when considering active surveillance versus more limited surgical resections, for example, a hemithyroidectomy. They found that an age of ≥ 55 years, hypoechogenicity, microcalcifications, irregular tumour shape, smooth margins, and high vascularity increased the risk for minimal extrathyroidal extension, lymph node metastasis, and capsular and vascular invasion [24]. It is certainly evident that some of these characteristics are very difficult to identify, both due to the small size of these lesions as well as due to the frequent lack of clinical suspicion. The findings of Slijepcevic et al. [25] are in agreement with Kaliszewski et al. [26] when it comes to the size of the lesion. They both mention that lesions under 5 mm are less likely to have intrathyroid extension and lymph node or distant metastases. The characteristics of imaging, and especially conventional ultrasound that is mentioned above (hypoechogenicity, microcalcifications, tumour shape, margins, and vascularity), are useful when aiming to categorise thyroid nodules, especially when attempting to identify malignant nodules. The Thyroid Imaging Reporting and Data System (TIRADS) has been developed and used widely based on such ultrasound characteristics. The TIRADS system is not particularly useful in "indeterminate nodules", for example, TIRADS III. Chen et al. [27] used conventional ultrasound-based radiomics to create a predictive model for identifying papillary thyroid microcarcinomas in patients with TIRADS III nodules. They created three predictive models: one based on radiomics, one using clinical features, and a predictive model combining both. They found excellent predictive results for papillary thyroid microcarcinoma in TIRADS III nodules. Interestingly, there was no statistical difference between the radiomics-based model and the combined radiomics-clinical-based model [27]. These results show promise for more accurate diagnostic techniques and better risk stratification of patients with benign thyroid disease and suspected papillary thyroid microcarcinomas.

In recent years, molecular profiling has become more feasible due to next-generation sequencing technology. Molecular profiling is the focus of many studies for different types of cancer. There is hope that these studies will aid the overall individualisation of risk profiling and cancer management in the future. Li et al. [28] performed whole exome and RNA sequencing on 64 patients with a papillary thyroid microcarcinoma (PTMC) and compared their findings to the patients in the Cancer Genome Atlas Program (TCGA). Apart from the well-known BRAF and RET mutations, they identified a molecular signature that they named PTMC-inflammatory, which they hypothesise presents a link to immune intervention [28]. Molecular profiling and further understanding of the immune microenvironment of these tumours can aid in the better classification of thyroid papillary microcarcinomas and risk stratification of patients. It can also provide potential targets for individualised treatment strategies, especially immunotherapy.

Papillary thyroid microcarcinomas are often multifocal (15.5–40% in surgical series and >80% in systematic autopsy studies) [1]. In the current study, 72.7% of PTMCs were multifocal, a percentage higher than reported in most surgical series. Kaliszewski et al. showed that multifocal or bilateral tumours and tumour sizes above 5 mm were the best predictors of lymph node metastasis in papillary thyroid microcarcinomas and thus require a more aggressive treatment approach; capsular invasion, age, and gender were not found to be statistically significant risk factors [26]. They also note that in cases of lymph node metastases or tumours larger than 5 mm, the entirety of thyroid tissue should be pathologically examined to exclude the multifocality of PTMCs [26]. This would result in a completion thyroidectomy in case a thyroid lobectomy was initially performed. Maturo et al. noted that the microcarcinoma foci were frequently found in the Zuckerkandl tubercle or in the pyramidal lobe, areas that are often not excised during thyroid lobectomy [29].

One of the main arguments used against a total thyroidectomy in benign thyroid disease is the increased risk for postoperative hypoparathyroidism as well as injury to recurrent laryngeal nerves. The postoperative complication rates reported in this study, 0.70% hypoparathyroidism and 0.94% unilateral vocal cord paresis, are similar to the postoperative complication rates reported in patients who underwent lobectomy [29,30].

Studies from specialist thyroid surgery units argue that the recurrent laryngeal nerve dissection technique, as well as thyroid vessel ligation near the thyroid capsule in total thyroidectomy by specialist thyroid surgeons, reduces the complication rates to levels comparable to lobectomy [29–31]. It needs to be stressed that a completion thyroidectomy in case of local recurrence post-lobectomy is certainly a more challenging operation with much higher complication rates [30].

The main limitation of this study is that it is a retrospective study from a single centre. The main benefit of the study, apart from the reasonable number of participants, is that all patients were operated by the same specialist endocrine surgical team and all specimens were examined in the same histopathology laboratory by histopathologists specialising in endocrine and thyroid pathology.

5. Conclusions

Papillary thyroid microcarcinomas are a common incidental finding in patients with benign thyroid disease, especially multinodular goitre. The incidental diagnosis of such lesions presents a challenge for the endocrine surgeon and can be especially stressful for the patient. It is important to consider all therapeutic options and tailor them to each patient based on their individual risk profiles in addition to the imaging characteristics and cytological characteristics of the microcarcinoma.

Based on the findings of this study, as well as a review of the literature, a total thyroidectomy should be considered as the operation of choice in patients with benign thyroid disease requiring surgical excision, especially in the case of multinodular goitre. Patient selection as well as open communication with the patient and the multidisciplinary team remain key in the management and follow-up of these patients.

There are multiple studies focusing on novel techniques for risk stratification and diagnosis of patients with papillary thyroid microcarcinomas. The improvement of imaging techniques, as well as the incorporation of computational methods and the development of radiomics, provide promise for more accurate diagnosis and individualisation of treatment approaches for patients with thyroid microcarcinoma. It should aid in providing the utmost diagnostic accuracy in order to avoid overtreatment without misdiagnosing in this particular patient group. Further prospective, randomised trials, as well as studies focusing on the molecular characteristics and immune microenvironment of papillary thyroid microcarcinomas, are needed in order to understand this distinct category of thyroid tumour and to determine the best treatment strategy for this increasingly common and diverse group of patients.

Author Contributions: Conceptualization, V.M. (Vasiliki Magra), T.K. and K.S.; methodology, V.M. (Vasiliki Magra), K.B., E.P., V.M. (Vasiliki Manaki) and C.S.; investigation, V.M. (Vasiliki Magra), I.M., K.B., E.P., V.M. (Vasiliki Manaki) and C.S.; collection and assembly of data, V.M. (Vasiliki Magra), I.M., K.B., E.P., V.M. (Vasiliki Manaki) and C.S.; writing—original draft preparation, all authors; writing—review and editing, all authors; supervision, K.S., I.K., T.K., S.M. and S.L.; project administration, K.S., T.K., I.K., T.K., S.M. and S.L. All authors have read and agreed to the published version of the manuscript.

Funding: This research received no external funding.

Institutional Review Board Statement: Not applicable.

Informed Consent Statement: Not applicable.

Data Availability Statement: Data are contained within the article.

Conflicts of Interest: The authors declare no conflicts of interest.

References

1. Dideban, S.; Abdollahi, A.; Meysamie, A.; Sedghi, S.; Shahriari, M. Thyroid Papillary Microcarcinoma: Etiology, Clinical Manifestations, Diagnosis, Follow-up, Histopathology and Prognosis. *Iran. J. Pathol.* **2016**, *11*, 1–19.
2. Boucek, J.; Kastner, J.; Skrivan, J.; Grosso, E.; Gibelli, B.; Giugliano, G.; Betka, J. Occult thyroid carcinoma. *Acta Otorhinolaryngol. Ital.* **2009**, *29*, 296–304.
3. Fama, F.; Sindoni, A.; Cicciu, M.; Polito, F.; Piquard, A.; Saint-Marc, O.; Gioffre-Florio, M.; Benvenga, S. Preoperatively undiagnosed papillary thyroid carcinoma in patients thyroidectomized for benign multinodular goiter. *Arch. Endocrinol. Metab.* **2018**, *62*, 139–148. [CrossRef]
4. Sobin, L.H.; Wittekind, C. *TNM Classification of Malignant Tumours*, 6th ed.; Wiley-Liss: New York, NY, USA, 2002.
5. Christofer Juhlin, C.; Mete, O.; Baloch, Z.W. The 2022 WHO classification of thyroid tumors: Novel concepts in nomenclature and grading. *Endocr. Relat. Cancer* **2022**, *30*, e220293. [CrossRef]
6. Mercante, G.; Frasoldati, A.; Pedroni, C.; Formisano, D.; Renna, L.; Piana, S.; Gardini, G.; Valcavi, R.; Barbieri, V. Prognostic Factors Affecting Neck Lymph Node Recurrence and Distant Metastasis in Papillary Microcarcinoma of the Thyroid: Results of a Study in 445 Patients. *Thyroid* **2009**, *19*, 707–716. [CrossRef]
7. Roti, E.; Rossi, R.; Trasforini, G.; Bertelli, F.; Ambrosio, M.R.; Busutti, L.; Pearce, E.N.; Braverman, L.E.; degli Uberti, E.C. Clinical and histological characteristics of papillary thyroid microcarcinoma: Results of a retrospective study in 243 patients. *J. Clin. Endocrinol. Metab.* **2006**, *91*, 2171–2178. [CrossRef] [PubMed]
8. Ardito, G.; Revelli, L.; Giustozzi, E.; Salvatori, M.; Fadda, G.; Ardito, F.; Avenia, N.; Ferretti, A.; Rampin, L.; Chondrogiannis, S.; et al. Aggressive papillary thyroid microcarcinoma: Prognostic factors and therapeutic strategy. *Clin. Nucl. Med.* **2013**, *38*, 25–28. [CrossRef] [PubMed]
9. Haugen, B.R.; Alexander, E.K.; Bible, K.C. 2015 American Thyroid Association Management Guidelines for Adult Patients with Thyroid Nodules and Differentiated Thyroid Cancer: The American Thyroid Association Guidelines Task Force on Thyroid Nodules and Differentiated Thyroid Cancer. *Thyroid* **2016**, *26*, 1–133. [CrossRef]
10. Spaziani, E.; Di Filippo, A.R.; Di Cristofano, C.; Tamagnini, G.T.; Spaziani, M.; Caruso, G.; Salina, G.; Valle, G.; Picchio, M.; De Cesare, A. Incidental papillary thyroid microcarcinoma in consecutive patients undergoing thyroid surgery for benign disease. A single center experience. *Ann. Ital. Chir.* **2023**, *94*, 142–146. [PubMed]
11. Carvalho, A.Y.; Kohler, H.F.; Gomes, C.C.; Vartanian, J.G.; Kowalski, L.P. Predictive Factors of Recurrence of Papillary Thyroid Microcarcinomas: Analysis of 2538 Patients. *Int. Arch. Otorhinolaryngol.* **2021**, *25*, e585–e593.
12. Lee, S.H.; Lee, S.S.; Jin, S.M.; Kim, J.H.; Rho, Y.S. Predictive Factors for Central Compartment Lymph Node Metastasis in Thyroid Papillary Microcarcinoma. *Laryngoscope* **2008**, *118*, 659–662. [CrossRef] [PubMed]
13. Bove, A.; Manunzio, R.; Palone, G.; Di Renzo, R.M.; Calabrese, G.V.; Perpetuini, D.; Barone, M.; Chiarini, S.; Mucilli, F. Incidence and Clinical Relevance of Incidental Papillary Carcinoma in Thyroidectomy for Multinodular Goiters. *J. Clin. Med.* **2023**, *12*, 2770. [CrossRef]
14. Hu, X.; Wang, X.; Liang, Y.; Chen, X.; Zhou, S.; Fei, W.; Yang, Y.; Que, H. Cancer Risk in Hashimoto's Thyroiditis: A Systematic Review and Meta-Analysis. *Front. Endocrinol.* **2022**, *13*, 937871. [CrossRef]
15. Unger, P.; Ewart, M.; Wang, B.Y.; Gan, L.; Kohtz, D.S.; Burstein, D.E. Expression of p63 in papillary thyroid carcinoma and in Hashimoto's thyroiditis: A pathobiologic link? *Hum. Pathol.* **2003**, *34*, 764–769. [CrossRef]
16. Jackson, D.; Handelsman, R.S.; Farrá, J.C.; Lew, J.I. Increased Incidental Thyroid Cancer in Patients with Subclinical Chronic Lymphocytic Thyroiditis. *J. Surg. Res.* **2020**, *245*, 115–118. [CrossRef]
17. Vargas-Uricoechea, H. Autoimmune Thyroid Disease and Differentiated Thyroid Carcinoma: A Review of the Mechanisms that Explain an Intriguing and Exciting Relationship. *World J. Oncol.* **2024**, *15*, 14–27. [CrossRef]
18. Lee, I.; Kim, H.K.; Soh, E.Y.; Lee, J. The Association Between Chronic Lymphocytic Thyroiditis and the Progress of Papillary Thyroid Cancer. *World J. Surg.* **2020**, *44*, 1506–1513. [CrossRef]
19. Lasithiotakis, K.; Grisbolaki, E.; Koutsomanolis, D.; Venianaki, M.; Petrakis, I.; Vrachassotakis, N.; Chrysos, E.; Zoras, O.; Chalkiadakis, G. Indications for surgery and significance of unrecognized cancer in endemic multinodular goiter. *World J. Surg.* **2012**, *36*, 1286–1292. [CrossRef]
20. Baier, N.D.; Hahn, P.F.; Gervais, D.A.; Samir, A.; Halpern, E.F.; Mueller, P.R.; Harisinghani, M.G. Fine-needle aspiration biopsy of thyroid nodules: Experience in a cohort of 944 patients. *AJR Am. J. Roentgenol.* **2009**, *193*, 1175–1179. [CrossRef] [PubMed]
21. Miccoli, P.; Minuto, M.N.; Galleri, D.; D'Agostino, J.; Basolo, F.; Antonangeli, L.; Aghini-Lombardi, F.; Berti, P. Incidental thyroid carcinoma in a large series of consecutive patients operated on for benign thyroid disease. *ANZ J. Surg.* **2006**, *76*, 123–126. [CrossRef] [PubMed]
22. Evranos, B.; Polat, S.B.; Cuhaci, F.N.; Baser, H.; Topaloglu, O.; Kilicarslan, A.; Kilic, M.; Ersoy, R.; Cakir, B. A cancer of undetermined significance: Incidental thyroid carcinoma. *Diagn. Cytopathol.* **2019**, *47*, 412–416. [CrossRef]
23. Lucandri, G.; Fiori, G.; Falbo, F.; Pende, V.; Farina, M.; Mazzocchi, P.; Santonati, A.; Bosco, D.; Spada, A.; Santoro, E. Papillary Thyroid Microcarcinoma: Differences between Lesions in Incidental and Nonincidental Settings-Considerations on These Clinical Entities and Personal Experience. *Curr. Oncol.* **2024**, *31*, 941–951. [CrossRef] [PubMed]
24. Kaliszewski, K.; Diakowska, D.; Rzeszutko, M.; Nowak, Ł.; Aporowicz, M.; Wojtczak, B.; Sutkowski, K.; Rudnicki, J. Risk factors of papillary thyroid microcarcinoma that predispose patients to local recurrence. *PLoS ONE* **2020**, *15*, e0244930. [CrossRef] [PubMed]

25. Slijepcevic, N.; Zivaljevic, V.; Diklic, A.; Jovanovic, M.; Oluic, B.; Paunovic, I. Risk factors associated with intrathyroid extension of thyroid microcarcinomas. *Langenbecks Arch. Surg.* **2018**, *403*, 615–622. [CrossRef]
26. Kaliszewski, K.; Diakowska, D.; Wojtczak, B.; Forkasiewicz, Z.; Pupka, D.; Nowak, L.; Rudnicki, J. Which papillary thyroid microcarcinoma should be treated as "true cancer" and which as "precancer"? *World J. Surg. Oncol.* **2019**, *17*, 91. [CrossRef]
27. Chen, Z.; Zhan, W.; Wu, Z.; He, H.; Wang, S.; Huang, X.; Xu, Z.; Yang, Y. The ultrasound-based radiomics-clinical machine learning model to predict papillary thyroid microcarcinoma in TI-RADS 3 nodules. *Transl. Cancer Res.* **2024**, *13*, 278–289. [CrossRef]
28. Li, Q.; Feng, T.; Zhu, T.; Zhang, W.; Qian, Y.; Zhang, H.; Zheng, X.; Li, D.; Yun, X.; Zhao, J.; et al. Multi-omics profiling of papillary thyroid microcarcinoma reveals different somatic mutations and a unique transcriptomic signature. *J. Transl. Med.* **2023**, *21*, 206. [CrossRef] [PubMed]
29. Maturo, A.; Tromba, L.; De Anna, L.; Carbotta, G.; Livadoti, G.; Donello, C.; Falbo, F.; Galiffa, G.; Esposito, A.; Biancucci, A.; et al. Incidental thyroid carcinomas. A retrospective study. *G. Chir. J. Ital. Surg. Assoc.* **2017**, *38*, 94–101. [CrossRef]
30. Sanabria, A.; Kowalski, L.P.; Tartaglia, F. Inferior thyroid artery ligation increases hypocalcemia after thyroidectomy: A meta-analysis. *Laryngoscope* **2018**, *128*, 534–541. [CrossRef]
31. Colombo, C.; De Leo, S.; Di Stefano, M.; Trevisan, M.; Moneta, C.; Vicentini, L.; Fugazzola, L. Total Thyroidectomy Versus Lobectomy for Thyroid Cancer: Single-Center Data and Literature Review. *Ann. Surg. Oncol.* **2021**, *28*, 4334–4344. [CrossRef]

Article

Safety and Feasibility of Single-Port Trans-Axillary Robotic Thyroidectomy: Experience through Consecutive 100 Cases

Il Ku Kang, Joonseon Park, Ja Seong Bae, Jeong Soo Kim and Kwangsoon Kim *

Department of Surgery, College of Medicine, The Catholic University of Korea, Seoul 06591, Korea
* Correspondence: noar99@naver.com; Tel.: +82-2-2258-6784; Fax: +82-2-2258-2138

Abstract: *Background and Objectives:* Recently, the single-port (SP) robotic system was introduced for minimally invasive operative techniques. Thus, this study aimed to validate the safety and feasibility of SP trans-axillary robotic thyroidectomy (SP-TART) through experiences in a single tertiary institution. *Materials and Methods:* This study retrospectively analyzed 100 consecutive patients who underwent SP-TART from October 2021 to June 2022 in Seoul St. Mary's Hospital in Seoul, Korea. We analyzed the clinicopathological characteristics and perioperative outcomes, including complications. *Results:* Less than total thyroidectomy (LTT) was performed in 81, total thyroidectomy (TT) in 16, and TT with modified radical neck dissection (mRND) in 3 patients. The mean operation time (min) was 53.3 ± 13.7, 86.3 ± 15.1, and 245.7 ± 36.7 in LTT, TT, and TT with mRND, respectively. The mean postoperative hospital stay was 2.0 ± 0.2, 2.1 ± 0.3, and 3.7 ± 1.5 days, respectively. A total of 84 cases of thyroid cancer were included, and 97.6% of them (82 cases) were papillary carcinoma and the rest were follicular and poorly differentiated carcinomas. Regarding complications, five cases had major complications, including three cases of vocal cord palsy and two cases of transient hypoparathyroidism. *Conclusions:* SP-TART is safe and feasible with a short operation time and a short length of hospital stay.

Keywords: robotic thyroidectomy; single-port robotic system; safety and feasibility; trans-axillary thyroidectomy

Citation: Kang, I.K.; Park, J.; Bae, J.S.; Kim, J.S.; Kim, K. Safety and Feasibility of Single-Port Trans-Axillary Robotic Thyroidectomy: Experience through Consecutive 100 Cases. *Medicina* **2022**, *58*, 1486. https://doi.org/10.3390/medicina58101486

Academic Editor: Jin Wook Yi

Received: 19 September 2022
Accepted: 17 October 2022
Published: 19 October 2022

Publisher's Note: MDPI stays neutral with regard to jurisdictional claims in published maps and institutional affiliations.

1. Introduction

Thyroid cancer is the most common cause of thyroidectomy. An estimated 43,800 adults will be diagnosed with thyroid cancer this year in the United States according to cancer statistics released by the American Cancer Society [1].

In the mid-2000s, robotic thyroidectomy was first introduced using a gasless trans-axillary approach [2,3]. Then, endocrine surgeons tried various types of remote access approaches, such as the bilateral axillo-breast, the gasless post-auricular facelift, and the trans-oral approach [4,5]. Trans-axillary robotic thyroidectomy (TART) is an application of an endoscopic thyroidectomy to a robotic system. Huscher et al. performed the first endoscopic thyroidectomy after Gagner pioneered the endoscopic parathyroidectomy [6,7]. Later, Chung et al. devised a gasless trans-axillary technique [8,9].

Conventional open thyroidectomy (COT) using Kocher's incision was long considered a safe operation among surgeons. However, COT is associated with two disadvantages. One is swallowing discomfort due to adhesion between the subplatysmal flap and the strap muscles [10]. Two is the scar location which is prone to exposure [11]. TART with a multi-port robotic system became widespread in Asian countries, where people care about other people's eyes or opinions because it can hide scars. Additionally, the short access distance from the armpit of Asian patients would have helped surgeons try TART in the early stage.

TART with the previous robotic system still has drawbacks [8,12], such as longer incisions and a broader extent of the skin flap than COT. Endoscopic thyroidectomy uses the same surgical instruments as in laparoscopic surgery, whereas TART uses three robotic arms with wrist articulation for counter-traction, dissection, and cutting with cauterization. However, the two methods require the same anatomical working space where an external retractor is inserted to lift the sternal head of the sternocleidomastoid muscle (SCM) and strap muscles surrounding the thyroid gland [2]. Therefore, some patients have postoperative pain and paresthesia in the supra- and infra-clavicular areas [13].

The single-port (SP) robotic system (Intuitive Surgical Inc., Sunnyvale, CA, USA) has advantages when it comes to thyroid surgery. SP trans-axillary robotic thyroidectomy (SP-TART) requires a shorter incision length and a narrower skin flap for working space that enables less sensory nerve injury [5]. However, little is known about its safety and feasibility. Thus, this study aimed to share experiences and help endocrine surgeons who would like to introduce SP-TART by reporting consecutively performed 100 cases in a single tertiary institution.

2. Materials and Methods

2.1. Patients

This study retrospectively analyzed 100 consecutive patients who underwent SP-TART from October 2021 to June 2022 in Seoul St. Mary's Hospital in Seoul, Korea. All patients who underwent SP-TART during the period were enrolled regardless of pathology or surgical extent. All operations were performed by a single surgeon (K.K.). The medical charts and pathology reports were reviewed and analyzed. There were no selection criteria for SP-TART, which was performed for all the patients who agreed with our procedure and possible complications.

This study was conducted following the Declaration of Helsinki (as revised in 2013) and approved by the Institutional Review Board of Seoul St. Mary's Hospital, the Catholic University of Korea (approval No.: KC22RISI0657), which waived the requirement for informed consent due to the retrospective nature of the study.

2.2. Operative Procedure of Less Than Total Thyroidectomy (LTT)

The patient is positioned supine with a soft pillow behind their shoulders to allow for neck extension under general anesthesia. The lesion-side arm is lifted overhead and fastened to expose the axilla. Landmarks, such as the cricoid cartilage, the sternal notch, and the upper pole of the ipsilateral lobe, are marked. A 3.5–4 cm curvilinear skin incision is made just posterior to the anterior axillary line so that the scar is covered when the arm is lowered. A subcutaneous skin flap is made along the anterior surface of the pectoralis major muscle and clavicle until the SCM is identified. The SCM is exposed after opening the superficial layer of the deep cervical fascia and raising the subplatysmal skin flap. The avascular plane between the sternal and clavicular heads of the SCM is identified and separated. The dissection is continued until the omohyoid, and other strap muscles are identified. The ipsilateral lobe of the thyroid gland is exposed after lifting the lateral border of the strap musculature upward and the internal jugular vein (IJV) downward. The space between the strap muscles and the thyroid gland is further dissected. After visualizing the upper pole of the ipsilateral lobe and part of the contralateral lobe of the thyroid gland, an external retractor [2] is inserted through the axillary skin incision to elevate the strap muscles. The procedure described above is based on the gasless method without using CO_2 insufflation during surgery.

The camera and instruments are inserted into the surgical field through a single cannula placed lateral to the skin incision. In the cannula, a camera is placed on the bottom; two pairs of Maryland dissecting forceps with bipolar electrocautery are positioned on both lateral-side arms, and a pair of Cadiere forceps are inserted on top.

All dissection and vessel ligations are performed via the Maryland bipolar dissectors. Intraoperative neuromonitoring is utilized for all patients. Thyroidectomy proceeded in the same manner as previous trans-axillary approaches with a single incision. The operator is seated at the console, and the assistant is placed next to the lesion-side of the patient.

The clavicular head of the SCM and the IJV are softly retracted downward with the suction device by the assistant. The ipsilateral lobe of the thyroid gland is retracted in the caudal direction. Dissection proceeded until the upper pole of the ipsilateral lobe is sufficiently mobilized. The superior thyroidal vessels are ligated in contact with the thyroid gland to preserve the external branch of the superior laryngeal nerve. The superior and inferior parathyroid glands are identified, dissected, and gently swept laterally while keeping the vasculature intact. Cadiere forceps are used to medially retract the ipsilateral lobe to better visualize the tracheoesophageal groove and the course of the recurrent laryngeal nerve (RLN). The perithyroidal tissue is dissected keeping the nerve from thermal injury by the energy device once the nerve is identified near the lower pole of the ipsilateral lobe. The ipsilateral lobe is carefully retracted and dissected from the trachea using a Maryland dissector. Care is taken to avoid thermal injury to the RLN at the ligament of Berry. The surgical specimen is removed through the axillary skin incision with a laparoscopic grasper. The Nerve Integrity Monitor (Medtronic Inc., Minneapolis, MN, USA) is utilized to confirm the RLN function. After the closed suction drain insertion, the axillary skin incision is closed with an absorbable subcuticular skin stapler (INSORB, Incisive Surgical Inc., Plymouth, MN, USA).

2.3. Operative Procedure of Total Thyroidectomy (TT)

The flap dissection for working space and the ipsilateral thyroid lobectomy is the same as the LTT procedure. However, an external retractor is inserted in a slightly upper direction than ipsilateral lobectomy due to the difficulty of approaching the contralateral upper pole of the thyroid gland. The contralateral thyroid lobectomy proceeds after taking out the specimen of the ipsilateral lobe of the thyroid gland of the neck.

The isthmic portion is first dissected and detached from the trachea to complete the thyroidectomy. Afterward, the upper pole of the contralateral lobe is dissected to be sufficiently mobilized. Care should be taken for the course of the RLN and the location of the parathyroid glands when dissection proceeds.

2.4. Operative Procedure of Modified Radical Neck Dissection (mRND)

The patient is placed in the supine position on the operating table with a soft bar below the shoulders under general anesthesia. The lesion-side arm is abducted to expose the axilla and lateral neck. A 7–10-cm skin incision is made in the axilla along the anterior axillary fold and the lateral border of the pectoralis major muscle. The flap is medially dissected over the SCM toward the midline of the anterior neck under the platysma muscle. Laterally, the trapezius muscle is identified and dissected upward along its anterior border. The mRND procedure is similar to that of the open-conventional procedure after the docking procedure with an external retractor. After finishing levels III, IV, and V node dissections, redocking is required to provide a better view for level II lymph node (LN) dissection. The external retractor is removed and reinserted through the axillary incision toward the submandibular gland superiorly. A drainage tube is inserted under the incision line after delivering the specimen. Finally, the wound is cosmetically closed [14].

2.5. Surgical Outcome Assessment

Surgical outcomes were evaluated based on the medical chart review of all patients to collect data on tumor size, multifocality, extrathyroidal extension (ETE), thyroiditis, number of harvested and metastatic LNs, amount of blood loss, postoperative hospital stay, and postoperative complications. The pathologic stage was classified according to the 8th edition of the American Joint Committee on Cancer/Union for International Cancer Control (AJCC/UICC) tumor-node-metastasis (TNM) staging system. The duration of surgical phases, including flap elevation, docking, console operation, and total operation times, was determined.

3. Results

3.1. Baseline Clinicopathological Characteristics of the Study Patients

Table 1 presents the clinicopathological characteristics of 100 cases in the study. The mean age was 41.8 ± 13.3 (range, 17–69) years, and 9 (9.0%) patients were male. The mean body mass index was 23.4 ± 3.6 (range, 16.5–32.8) kg/m^2. LTT was performed in 81 (81.0%) patients, which included lobectomy with tumerectomy, isthmusectomy, and simple lobectomy. One patient was confirmed to have a poorly differentiated thyroid carcinoma (PDTC) after lobectomy and underwent completion of TT. Additionally, 16 (16.0%) patients underwent TT, including this case, while TT with mRND was performed in 3 (3.0%) patients. In terms of pathology, papillary carcinoma (PTC) accounted for 82%, while follicular adenoma and noninvasive follicular neoplasm with papillary-like nuclear features were 8% and 3%, respectively. The mean tumor size was 1.5 ± 1.3 (range, 0.3–7.0) cm and 1.3 ± 1.2 (range, 0.1–5.9) cm, when measured in ultrasound and pathology, respectively. Thyroiditis was confirmed in 46 (46.0%) cases.

Table 1. Baseline clinicopathological characteristics of the study patients.

	Total 100 Cases
Age (years)	41.8 ± 13.3 (range, 17–69)
Male:Female	1: 10.1 (9:91)
Body mass index (kg/m^2)	23.4 ± 3.6 (range, 16.5–32.8)
Extent of operation	
LTT	81 (81.0%)
TT	16 (16.0%)
TT with mRND	3 (3.0%)
Direction of approach	
Right/Left	56 (56.0%)/44 (44.0%)
Pathologic subtype	
PTC	82 (82.0%)
Follicular adenoma	8 (8.0%)
NIFTP	3 (3.0%)
FTC	1 (1.0%)
Nodular Hashimoto thyroiditis	1 (1.0%)
Oncocytic (Hurthle cell) adenoma	1 (1.0%)
Nodular hyperplasia with oncocytic (Hurthle cell) changes	1 (1.0%)
PDTC	1 (1.0%)
Diffuse hyperplasia with Hashimoto thyroiditis	1 (1.0%)
No residual tumor (completion TT) *	1 (1.0%)
Tumor size (cm) **	
Sonographic (n = 98)	1.5 ± 1.3 (range, 0.3–7.0)
Pathologic (n = 97) ***	1.3 ± 1.2 (range, 0.1–5.9)
Thyroiditis	
Hashimoto/Focal	40 (40.0%)/6 (6.0%)

Data are expressed as the patient number (%) or mean $\pm$ SD. * The patient with PDTC underwent two surgeries to perform completion TT. ** The tumor size was measured based on the largest one when multifocal tumors were observed. *** The tumor size was not reported as measured in the pathology of one patient with PTC. Abbreviations: LTT, less than total thyroidectomy; TT, total thyroidectomy; mRND, modified radical neck dissection; PTC, papillary thyroid carcinoma; NIFTP, noninvasive follicular neoplasm with papillary-like nuclear features; FTC, follicular thyroid carcinoma; PDTC, poorly differentiated thyroid carcinoma.

3.2. Perioperative Outcomes According to the Extent of Operation

Table 2 summarizes the perioperative outcomes according to the extent of the operation. the mean operation time (min) was 53.3 ± 13.7, 86.3 ± 15.1, and 245.7 ± 36.7, and the average amount of blood loss (ml) was 8.4 ± 5.4, 23.3 ± 48.9, and 38.3 ± 53.5 in LTT, TT, and TT with mRND, respectively. The mean postoperative hospital stay was 2.0 ± 0.2, 2.1 ± 0.3, and 3.7 ± 1.5 days, respectively. Out of 9 patients who had postoperative complications, 6 (7.4%) underwent LTT, 2 (12.5%) underwent TT, and 1 (33.3%) underwent TT with mRND.

Table 2. Perioperative outcomes according to the extent of operation.

	LTT (n = 81)	TT (n = 16)	TT with mRND (n = 3)
Operation time (min)	53.3 ± 13.7	86.3 ± 15.1	245.7 ± 36.7
Flap time	14.8 ± 2.8	22.8 ± 3.6	39.0 ± 5.6
Docking time	2.1 ± 0.9	2.0 ± 0.0	4.0 ± 0.0
Console time	22.5 ± 9.9	47.8 ± 12.2	158.0 ± 14.4
Blood loss (ml)	8.4 ± 5.4	23.3 ± 48.9	38.3 ± 53.5
Hospital stay (POD)	2.0 ± 0.2	2.1 ± 0.3	3.7 ± 1.5
Postoperative complications	6 (7.4%)	2 (12.5%)	1 (33.3%)

Data are expressed as the patient number (%) or mean ± SD. Abbreviations: LTT, less than total thyroidectomy; TT, bilateral total thyroidectomy; mRND, modified radical neck dissection; POD, postoperative day.

3.3. Clinicopathological Characteristics of Patients with Cancer

The clinicopathological characteristics of patients with cancer are demonstrated in Table 3. On a pathological basis, 84 cases were cancer. Additionally, 82 (97.6%) cases were PTC, while the others were 1 (1.2%) follicular carcinoma and 1 (1.2%) PDTC. Surgical extent accounted for 79.8%, 20.2%, and 3.6% for LTT, TT, and TT with mRND, respectively. The mean tumor size was 1.3 ± 1.3 (range, 0.3–7.0) cm and 1.1 ± 1.2 (range, 0.1–5.9) cm, when measured in ultrasound and pathology, respectively. Multifocal tumors were observed in 30 (35.7%) cases. Tumors were invaded to the outside of the capsule of the thyroid gland in approximately two-thirds of the patients with cancer (55 cases, 65.5%), of which 5 (6.0%) cases had gross ETE. Thyroiditis was pathologically confirmed in half of the patients with cancer (42/84 cases). In terms of 81 patients (96.4% of cancer cases) who underwent LN dissection, the mean harvested LNs was 8.0 ± 9.0 (range, 0–56), while LNs positive for metastasis were 2.4 ± 4.8 (range, 0–27). According to the eighth version of the AJCC/UICC TNM staging system, 79 (94.0%) patients were pathological stage I, and 5 (6.0%) were stage II.

Table 3. Clinicopathological characteristics of the patients with cancer.

	Total 84 Cases
Extent of operation	
LTT/TT/mRND	67 (79.8%)/14 (16.6%)/3 (3.6%)
Pathologic subtype	
PTC	82 (97.6%)
FTC	1 (1.2%)
PDTC	1 (1.2%)
Tumor size (cm)	
Sonographic	1.3 ± 1.3 (range, 0.3–7.0)
Pathologic	1.1 ± 1.2 (range, 0.1–5.9)
Multifocality	30 (35.7%)
ETE	
Minimal/Gross	50 (59.5%)/5 (6.0%)
Thyroiditis	
Hashimoto/Focal	36 (42.9%)/6 (7.1%)
LN dissection	81 (96.4%)
Harvested LNs	8.0 ± 9.0 (range, 0–56)
Positive LNs	2.4 ± 4.8 (range, 0–27)
T stage	
T1a/T1b/T2/T3a/T3b	59 (70.2%)/12 (14.2%)/3 (3.6%)/5 (6.0%)/5 (6.0%)
N stage	
N0/N1a/N1b/Nx	38 (45.2%)/39 (46.4%)/3 (3.6%)/4 (4.8%)
TNM stage	
Stage I/II	79 (94.0%)/5 (6.0%)

Data are expressed as the patient number (%) or mean ± SD. Abbreviations: LTT, less than total thyroidectomy; TT, bilateral total thyroidectomy; mRND, modified radical neck dissection; PTC, papillary thyroid carcinoma; FTC, follicular thyroid carcinoma; PDTC, poorly differentiated thyroid carcinoma; ETE, extrathyroidal extension; LN, lymph node; T, tumor; N, node; M, metastasis; TNM, tumor-node-metastasis.

3.4. Postoperative Complications of the Study Participants

Postoperative complications are listed in Table 4. Nine patients experienced postoperative complications, including three vocal cord palsy, two surgical site infections, and two transient hypoparathyroidisms. Additionally, chyle leakage, bleeding at the drain insertion site, and wound dehiscence occurred in one case each. Of the patients with vocal cord palsy, two patients developed after LTT while one patient was after TT. A case of intraoperative transection of the recurrent laryngeal nerve was developed in one patient during left thyroid lobectomy.

Table 4. Postoperative complications of the study participants.

	Total 100 Cases
Vocal cord palsy	3 (3.0%)
Surgical site infection	2 (2.0%)
Transient hypoparathyroidism *	2 (2.0%)
Chyle leakage **	1 (1.0%)
Drain insertion site bleeding	1 (1.0%)
Wound dehiscence	1 (1.0%)

Data are expressed as the patient number (%). * One was developed after TT while the other was after TT with mRND. ** The same patient experienced chyle leakage and transient hypoparathyroidism after TT with mRND, left. Abbreviations: TT, bilateral total thyroidectomy; mRND, modified radical neck dissection.

4. Discussion

The concept of endocrine surgery has changed with time. In the past, surgeons were focused on the organ that had to be removed. Some patients suffered from voice change or hypocalcemia postoperatively. Nowadays, both surgeons and patients are interested in preserving the function of adjacent structures as much as possible. Since Gagner published a paper on the first endoscopic subtotal parathyroidectomy, many have developed remote-access thyroid surgery with various approaches [6,7]. However, the experience of TART is limited to a few countries, such as South Korea [15]. Concerns about different surgical outcomes between TART and COT remained [16]. We analyzed 100 consecutive patients who underwent SP-TART at a single tertiary institution. Herein, we would like to discuss our learning from our experience.

SP robotic system has a single arm with a flexible scope and three multi-jointed instruments that can easily approach blind spots without preparing a broad flap dissection [17,18]. The camera and instruments in the previous multi-port system have no flexibility, but it was multiple arms connected with them that moved. In comparison, the latest flexible system has instruments with multiple joints in the distal part to move like a fiber-optic endoscope. Making a long vertical incision and creating a broad flap dissection along the pectoralis major muscle and the clavicle was not necessary because the proximal part of the instruments and the camera are rigid and clustered like a bundle in a single cannula. The three-dimensional high-resolution camera is fully wristed to reach narrow spaces or blind spots. With the high resolution of the camera and the counter-traction of the third arm, differentiating the parathyroid glands and finding the course of the RLN was much easier in TART than in COT.

Kim et al. studied 200 cases of SP-TART and revealed that the position of the patient's arm was laterally extended during upper pole dissection [18]. In contrast, we never had the patient's arm moved during surgery. All patients were kept in the same position during all surgical procedures. The lesion-side arm was fixed overhead so that the armpit could be as close to the neck as possible for those who underwent LTT and TT. Patients who underwent TT with mRND were positioned with their lesion-side arm laterally extended.

A 3.5–4-cm long incision was made for LTT and TT, while a 7–8-cm long incision for TT with mRND. Based on our experience, relatively short incisions and a narrow extent of flap dissection were needed for SP-TART compared to the previous robotic system. A 3.5–4 cm long incision was sufficient to make a working space and for the instruments and camera to move, regardless of tumor size. Despite the smaller incision size, putting gauze in or taking the specimen out was not as difficult for the assistant as the previous multi-port robotic system. Less collision with the instruments occurred in the SP robot setting.

The operative view of TART is a lateral view which is similar to COT. Manipulating the upper pole of the ipsilateral lobe and dissecting the central LNs is relatively easier compared with the bilateral axillo-breast or trans-oral approach [2,19]. However, the upper pole of the contralateral lobe is more difficult to approach when performing TT. In this study, the console time was 22.5 ± 9.9 min in the LTT group (n = 81), while 47.8 ± 12.2 min in the TT group (n = 16), which was more than twice as long. This time difference will be shorter, as the surgeon gets more experienced.

The absence of neck incision postoperatively is one of the greatest benefits of SP-TART. The normal arm position conceals the surgical scar in the axilla, providing good cosmetic results [11]. Most of the patients undergoing thyroid surgery are females and are under the age of 40 years. In our data, the mean age of 100 patients was 41.8 years and 91 of them were female. Some young females, especially those who are interested in beauty, are reluctant to expose surgical wounds. Endocrine surgeons need to cater to those needs. The axillary approach has the advantage of avoiding swallowing symptoms postoperatively in addition to cosmetic outcomes [11]. Several studies revealed that TART is safe and feasible and not inferior to COT [19–22].

SP-TART is a safe operation regardless of the patient's physique. Creating a working space via the axilla is easier in female patients with short height than in tall males or obese patients due to the difference in the extent of the skin flap [23]. However, this is not difficult for an experienced surgeon to overcome. Data from Yonsei University in South Korea revealed that obesity was not a major factor that affect the oncological and surgical outcomes of TART [24]. This study revealed that 6 patients have a body mass index of >30 kg/m^2, and one of them experienced vocal cord palsy postoperatively due to intraoperative transection of the RLN. Making trans-axillary skin flaps is considered more difficult in male patients due to the long distance and tightness. Our data included 7 male patients, of whom two had postoperative complications. One experienced vocal cord palsy after TT and the other experienced transient hypoparathyroidism and chyle leakage after TT with mRND.

Generally, TART does not differ in the oncological outcome of low-risk groups with early thyroid cancer compared to COT [2,12,25–27]. The demand for TART increased as the number of patients with thyroid microcarcinoma increased in the 2010's due to the development of ultrasound images in quality. Furthermore, TART was performed for advanced patients who needed mRND. The 5-year surgical and oncological outcomes between open mRND and robot mRND groups were not significantly different [28].

This study included three patients who underwent TT with ipsilateral mRND. Dissecting LNs at the upper jugular level was easier with SP robotic system than the previous robotic system developed for multi-port surgery. All the procedures, including both thyroid glands and supraclavicular LN dissection, were performed with the patient's arm laterally extended. Afterward, de-docking and redocking were needed for more flap dissection to create a working space. The patient's head was rotated toward the contralateral lobe and the external retractor [18] was reinserted so that the tip of the retractor could reach the upper jugular level of the ipsilateral lateral neck. We learned that additional retraction by an assistant was not necessary when dissecting the LNs at level II with the help of the flexible instruments and camera.

Studies on postoperative pain of TART were reported. Prospective studies reported that TART did not cause less postoperative pain than COT [13,29]. Similarly, we observed that quite a few patients had a hard time with postoperative pain at the location of subcutaneous flap dissection after SP-TART. We performed ultrasonography-guided pectoralis nerve block II before making an incision for some patients undergoing SP-TART (n = 28) and compared the visual analog scale with the no-block group (n = 27). The block group had significantly lower pain scores than the no-block group [30]. SP-TART may cause pain shortly postoperatively, but it can be overcome by a pectoralis nerve block.

One of the shortcomings of TART is sensory nerve injury caused by subcutaneous flap dissection [5]. Some patients appeal to paresthesia on the supra- and infra-clavicular region for several months postoperatively. The flexible SP robotic system enabled less extent of flap dissection which led to minimizing the sensory nerve injury.

This study has limitations. One is its retrospective nature. Additionally, there is something to consider as selection bias in this study. Getting enough profits from treatment covered by insurance is difficult in South Korea. Robotic surgery is not covered by public health insurance; thus, it has been facilitated for the last decade. Some patients can afford to pay private insurance, while others cannot. Therefore, there may be more demand for robotic surgery in certain regions with better socioeconomic status.

Of the total 100 cases in this study, 10 were complicated, and 5 of them were associated with the RLN and the parathyroid gland. Bleeding occurred in one case. No case was converted to open surgery intraoperatively, and there was no re-operation case postoperatively. Considering the small sample size of this study, SP-TART is safe and feasible with a low complication rate and short postoperative hospital stay. The complication rate will decrease as the surgical team gets more experience.

5. Conclusions

To our best knowledge, SP-TART is safe and technically feasible with a short incision length, a short hospital stay, and a relatively low complication rate. However, further prospective studies are needed to verify the technical feasibility and evaluate the operative outcomes. We expect this study to help endocrine surgeons who would like to perform SP-TART.

Author Contributions: Conceptualization: K.K.; methodology: K.K., J.S.B. and J.S.K.; software: K.K., J.P. and I.K.K.; validation: K.K., J.P. and I.K.K.; formal analysis: K.K. and I.K.K.; investigation: K.K. and I.K.K.; resources: K.K., J.S.B. and J.S.K.; data curation: K.K. and I.K.K.; writing–original draft preparation: K.K. and I.K.K.; writing–review and editing: all authors; visualization: K.K. and I.K.K.; supervision: J.S.B. and J.S.K.; project administration: K.K. All authors have read and agreed to the published version of the manuscript.

Funding: This research received no external funding.

Institutional Review Board Statement: This preliminary, retrospective cohort study was conducted in accordance with the Declaration of Helsinki (2013) and approved by the Institutional Review Board of Seoul St. Mary's Hospital, Catholic University of Korea (Seoul, Korea) (approval number:); the board waived the requirement for informed consent given the retrospective nature of the work.

Informed Consent Statement: Patient consent was waived due to the retrospective nature of this study.

Data Availability Statement: The data underlying this article cannot be shared publicly to maintain the privacy of individuals that participated in the study. The data will be shared upon reasonable request to the corresponding author.

Acknowledgments: We would like to thank all the nurses who participated in the surgery and contributed to this study.

Conflicts of Interest: The authors declare no conflict of interest.

References

1. Siegel, R.L.; Miller, K.D.; Fuchs, H.E.; Jemal, A. Cancer statistics, 2022. *CA Cancer J. Clin.* **2022**, *72*, 7–33. [CrossRef] [PubMed]
2. Kang, S.W.; Jeong, J.J.; Yun, J.S.; Sung, T.Y.; Lee, S.C.; Lee, Y.S.; Nam, K.H.; Chang, H.S.; Chung, W.Y.; Park, C.S. Robot-assisted endoscopic surgery for thyroid cancer: Experience with the first 100 patients. *Surg. Endosc.* **2009**, *23*, 2399–2406. [CrossRef]
3. Lobe, T.E.; Wright, S.K.; Irish, M.S. Novel uses of surgical robotics in head and neck surgery. *J. Laparoendosc. Adv. Surg. Tech. A* **2005**, *15*, 647–652. [CrossRef] [PubMed]
4. Tae, K. Robotic thyroid surgery. *Auris Nasus Larynx* **2021**, *48*, 331–338. [CrossRef]
5. Chang, E.H.E.; Kim, H.Y.; Koh, Y.W.; Chung, W.Y. Overview of robotic thyroidectomy. *Gland Surg.* **2017**, *6*, 218–228. [CrossRef]
6. Hüscher, C.S.; Chiodini, S.; Napolitano, C.; Recher, A. Endoscopic right thyroid lobectomy. *Surg. Endosc.* **1997**, *11*, 877.

7. Gagner, M. Endoscopic subtotal parathyroidectomy in patients with primary hyperparathyroidism. *Br. J. Surg.* **1996**, *83*, 875. [CrossRef] [PubMed]
8. Yoon, J.H.; Park, C.H., Chung, W.Y. Gasless endoscopic thyroidectomy via an axillary approach: Experience of 30 cases. *Surg. Laparosc. Endosc. Percutan. Tech.* **2006**, *16*, 226–231. [CrossRef] [PubMed]
9. Tae, K.; Ji, Y.B.; Song, C.M.; Ryu, J. Robotic and endoscopic thyroid surgery: Evolution and advances. *Clin. Exp. Otorhinolaryngol.* **2019**, *12*, 1–11. [CrossRef]
10. Hyun, K.; Byon, W.; Park, H.J.; Park, Y.; Park, C.; Yun, J.S. Comparison of swallowing disorder following gasless transaxillary endoscopic thyroidectomy versus conventional open thyroidectomy. *Surg. Endosc.* **2014**, *28*, 1914–1920. [CrossRef]
11. Lee, J.; Nah, K.Y.; Kim, R.M.; Ahn, Y.H.; Soh, E.Y.; Chung, W.Y. Differences in postoperative outcomes, function, and cosmesis: Open versus robotic thyroidectomy. *Surg. Endosc.* **2010**, *24*, 3186–3194. [CrossRef] [PubMed]
12. Kang, S.W.; Park, J.H.; Jeong, J.S.; Lee, C.R.; Park, S.; Lee, S.H.; Jeong, J.J.; Nam, K.H.; Chung, W.Y.; Park, C.S. Prospects of robotic thyroidectomy using a gasless, transaxillary approach for the management of thyroid carcinoma. *Surg. Laparosc. Endosc. Percutan. Tech.* **2011**, *21*, 223–229. [CrossRef] [PubMed]
13. Fregoli, L.; Materazzi, G.; Miccoli, M.; Papini, P.; Guarino, G.; Wu, H.S.; Miccoli, P. Postoperative pain evaluation after robotic transaxillary thyroidectomy versus conventional thyroidectomy: A prospective study. *J. Laparoendosc. Adv. Surg. Tech. A* **2017**, *27*, 146–150. [CrossRef] [PubMed]
14. Kim, K.; Kang, S.-W.; Chung, W.Y. Robotic-assisted modified radical neck dissection: Transaxillary, bilateral axill-breast approach (BABA), Facelift. *Curr. Surg. Rep.* **2019**, *7*, 20. [CrossRef]
15. Materazzi, G.; Fregoli, L.; Papini, P.; Bakkar, S.; Vasquez, M.C.; Miccoli, P. Robot-assisted transaxillary thyroidectomy (RATT): A series appraisal of more than 250 cases from Europe. *World J. Surg.* **2018**, *42*, 1018–1023. [CrossRef]
16. Jacobs, D.; Torabi, S.J.; Gibson, C.; Rahmati, R.; Mehra, S.; Judson, B.L. Assessing national utilization trends and outcomes of robotic and endoscopic thyroidectomy in the United States. *Otolaryngol. Head Neck Surg.* **2020**, *163*, 947–955. [CrossRef]
17. Kim, K.; Kang, S.W.; Kim, J.K.; Lee, C.R.; Lee, J.; Jeong, J.J.; Nam, K.H.; Chung, W.Y. Robotic transaxillary hemithyroidectomy using the da VINCI SP robotic system: Initial experience with 10 consecutive cases. *Surg. Innov.* **2020**, *27*, 256–264. [CrossRef]
18. Kim, J.K.; Choi, S.H.; Choi, S.M.; Choi, H.R.; Lee, C.R.; Kang, S.W.; Jeong, J.J.; Nam, K.H.; Chung, W.Y. Single-port transaxillary robotic thyroidectomy (START): 200-cases with two-step retraction method. *Surg. Endosc.* **2022**, *36*, 2688–2696. [CrossRef]
19. Kang, S.W.; Lee, S.C.; Lee, S.H.; Lee, K.Y.; Jeong, J.J.; Lee, Y.S.; Nam, K.H.; Chang, H.S.; Chung, W.Y.; Park, C.S. Robotic thyroid surgery using a gasless, transaxillary approach and the da VINCI S system: The operative outcomes of 338 consecutive patients. *Surgery* **2009**, *146*, 1048–1055. [CrossRef]
20. Kandil, E.H.; Noureldine, S.I.; Yao, L.; Slakey, D.P. Robotic transaxillary thyroidectomy: An examination of the first one hundred cases. *J. Am. Coll. Surg.* **2012**, *214*, 558–564, discussion 64–66. [CrossRef]
21. Sun, G.H.; Peress, L.; Pynnonen, M.A. Systematic review and meta-analysis of robotic vs conventional thyroidectomy approaches for thyroid disease. *Otolaryngol. Head Neck Surg.* **2014**, *150*, 520–532. [CrossRef] [PubMed]
22. de Vries, L.H.; Aykan, D.; Lodewijk, L.; Damen, J.A.A.; Rinkes, I.H.M.B.; Vriens, M.R. Outcomes of minimally invasive thyroid surgery—A systematic review and meta-analysis. *Front. Endocrinol.* **2021**, *12*, 719397. [CrossRef]
23. Holsinger, F.C.; Chung, W.Y. Robotic thyroidectomy. *Otolaryngol. Clin. N. Am.* **2014**, *47*, 373–378. [CrossRef] [PubMed]
24. Yap, Z.; Kim, W.W.; Kang, S.W.; Lee, C.R.; Lee, J.; Jeong, J.J.; Nam, K.H.; Chung, W.Y. Impact of body mass index on robotic transaxillary thyroidectomy. *Sci. Rep.* **2019**, *9*, 8955. [CrossRef] [PubMed]
25. Kang, S.W.; Jeong, J.J.; Nam, K.H.; Chang, H.S.; Chung, W.Y.; Park, C.S. Robot-assisted endoscopic thyroidectomy for thyroid malignancies using a gasless transaxillary approach. *J. Am. Coll. Surg.* **2009**, *209*, e1–e7. [CrossRef] [PubMed]
26. Lee, S.; Ryu, H.R.; Park, J.H.; Kim, K.H.; Kang, S.W.; Jeong, J.J.; Nam, K.H.; Chung, W.Y.; Park, C.S. Excellence in robotic thyroid surgery: A comparative study of robot-assisted versus conventional endoscopic thyroidectomy in papillary thyroid microcarcinoma patients. *Ann. Surg.* **2011**, *253*, 1060–1066. [CrossRef]
27. Tae, K.; Song, C.M.; Ji, Y.B.; Sung, E.S.; Jeong, J.H.; Kim, D.S. Oncologic outcomes of robotic thyroidectomy: 5-year experience with propensity score matching. *Surg. Endosc.* **2016**, *30*, 4785–4792. [CrossRef]
28. Kim, M.J.; Lee, J.; Lee, S.G.; Choi, J.B.; Kim, T.H.; Ban, E.J.; Lee, C.R.; Kang, S.W.; Jeong, J.J.; Nam, K.H.; et al. Transaxillary robotic modified radical neck dissection: A 5-year assessment of operative and oncologic outcomes. *Surg. Endosc.* **2017**, *31*, 1599–1606. [CrossRef]
29. Ryu, H.R.; Lee, J.; Park, J.H.; Kang, S.W.; Jeong, J.J.; Hong, J.Y.; Chung, W.Y. A comparison of postoperative pain after conventional open thyroidectomy and transaxillary single-incision robotic thyroidectomy: A prospective study. *Ann. Surg. Oncol.* **2013**, *20*, 2279–2284. [CrossRef]
30. Chae, M.S.; Park, Y.; Shim, J.W.; Hong, S.H.; Park, J.; Kang, I.K.; Bae, J.S.; Kim, J.S.; Kim, K. Clinical application of pectoralis nerve block II for flap dissection-related pain control after robot-assisted transaxillary thyroidectomy: A preliminary retrospective cohort study. *Cancers* **2022**, *14*, 4097. [CrossRef]

Article

Surgical Outcomes of Thyroidectomy in Geriatric Patients Aged 80 Years and Older: A Single-Center Retrospective Cohort Study

Wei Huang, Yi-Ju Chen and Wei-Hsin Chen *

Division of General Surgery, Department of Surgery, Taichung Veterans General Hospital, Taichung 407219, Taiwan; j300521@gmail.com (W.H.)
* Correspondence: maxchen0127@yahoo.com.tw

Abstract: *Background and Objectives*: As the global aging population grows, the incidence of thyroidectomy in elderly patients is increasing. This study aimed to evaluate the surgical outcomes of thyroidectomy in patients aged 80 years and older. *Materials and Methods*: All patients aged 80 years and older who underwent thyroidectomies at our hospital between January 2015 and December 2022 were reviewed in this retrospective cohort study. Collected data consisted of patients' clinical characteristics, functional status, compression symptoms, preoperative assessments, perioperative outcomes, postoperative complications (such as bleeding events, recurrent laryngeal nerve injury, hypocalcemia), pathological findings, readmission, and follow-up outcomes. *Results*: Seventeen patients were included in this study, with female predominance (82.4%). The mean age was 85.6 ± 4.8 years. Fourteen patients (82.4%) exhibited compression-related symptoms as surgical indications. Based on pathological reports, patients were categorized into benign (12/17, 70.6%) and malignancy (5/17, 29.4%) groups. The benign group had a shorter operation time compared with the malignancy group (164.3 ± 32.0 min vs. 231.0 ± 79.1 min, *p* = 0.048). No major postoperative complications developed. The median postoperative follow-up duration was 28 months (range: 2–91 months). Thirteen patients (76.5%) were alive at the end of the study period. *Conclusions*: Despite potential age-related risks, thyroidectomy is feasible for carefully selected patients aged 80 years and older. It provides benefits not only in terms of oncological curative treatment but also in improving the quality of life, such as compressive symptoms and wound condition.

Keywords: frailty; geriatric; multinodular goiter; thyroid cancer; thyroidectomy

Citation: Huang, W.; Chen, Y.-J.; Chen, W.-H. Surgical Outcomes of Thyroidectomy in Geriatric Patients Aged 80 Years and Older: A Single-Center Retrospective Cohort Study. *Medicina* **2024**, *60*, 1383. https://doi.org/10.3390/medicina60091383

Academic Editor: Jin Wook Yi

Received: 19 July 2024
Revised: 20 August 2024
Accepted: 22 August 2024
Published: 23 August 2024

1. Introduction

As disease prevention and medical advancements continue, physicians will be required to provide care to the growing older patient population. In 2020, 9% of the global population was aged over 65 years old. According to the population projection report of Taiwan, it is estimated that by 2025, the country will have a super-aged society, with individuals aged over 65 accounting for more than 20% of the total population [1]. The global older population is expected to increase more than twofold in 2050 [2].

Thyroid nodules represent a prevalent condition in the general population, with a prevalence increasing with age [3]. Older adults also have a higher prevalence of high-risk thyroid cancer, as well as an increased risk of multimorbidity, functional decline, and postoperative complications [3,4].

As society continues to age, the proportion of older patients with thyroid tumors also increases. Understanding the risks and outcomes associated with thyroid surgery in these patients is crucial.

Thyroid surgery in older patients is mainly performed in cases suspected of malignancy, thyrotoxicosis, or compressive symptoms. However, patients with difficulty in breathing or shortness of breath often first seek examination by a cardiologist or pulmonologist. This can inadvertently contribute to diagnostic misdirection, as the potential underlying thyroid disorders may remain unrecognized.

Clinically, we often encounter patients or their families expressing concerns about the limited remaining lifespan, which leads to hesitancy regarding surgical interventions for those over 80 years. Therefore, to ensure that our findings are more applicable to the clinical setting, we chose a cutoff age of 80 years for our analysis. Additionally, the existing data are limited and have shown conflicting results, and data on Asian populations are lacking [5–10]. Therefore, we aimed to evaluate the detailed characteristics and outcomes of patients aged 80 years and older who underwent thyroid surgery.

2. Materials and Methods

2.1. Patient Population and Study Design

In this retrospective observational cohort study, patients who underwent thyroid-related surgery at our hospital between January 2015 and December 2022 were identified using the operation schedule system in the department's database. Surgery for unilateral and bilateral total thyroidectomy was included. Patients were included if they were aged 80 years and older while undergoing thyroidectomy. A total of nineteen patients were included in the study. No patients with anaplastic carcinoma that underwent surgery were identified. After reviewing their medical records, two patients with thyroid cancer recurrence and those who underwent neck lymph node dissection without thyroidectomy were excluded. Subsequently, we selected patients with malignant pathological reports for further data analysis. This study ended in June 2023.

2.2. Thyroidectomy Protocol

All patients underwent preoperative neck ultrasonography. Patients identified by the anesthesiology department as having higher surgical risks (such as American Society of Anesthesiologists [ASA] classification over class 2 and history of heart disease) were scheduled for preoperative echocardiography to assess cardiac function. If the preoperative electrocardiogram and echocardiography showed any abnormalities, the patient was referred to a cardiovascular specialist for further evaluation. In cases where percutaneous coronary intervention or stent placement was needed, the operation was delayed according to guideline recommendations. All patients' blood pressure was typically managed to meet the standards of maintaining systolic blood pressure (SBP) below 180 mmHg and diastolic blood pressure (DBP) below 110 mmHg before surgery.

Patients presenting with airway symptoms, such as wheezing or difficulty in breathing, were scheduled for a pulmonary function test to assess lung capacity and function.

All thyroidectomies were performed by each of the two experienced endocrine surgeons in our department. One attending surgeon has 13 years of experience, with a thyroidectomy volume of approximately 250 patients annually and treated 9 patients in this study. The other surgeon has 15 years of experience, with a thyroidectomy volume of approximately 180 and treated 8 patients in this study. During surgery, intraoperative neurophysiological monitoring (IONM) equipment and an energy device (HARMONIC® scalpel, Ethicon, Cincinnati, OH, USA, or LigaSure™ sealer, Valleylab, Boulder, CO, USA) were used routinely, unless the patient could not afford the cost.

On the day after total thyroidectomy, both surgeons checked serum total calcium concentration to see if signs and symptoms of hypocalcemia were observed. One surgeon routinely monitored serum total calcium concentration on operation day and postoperative day 1, whereas the other surgeon did so only in response to symptomatic patients. Prophylactic calcium supplementation was not routinely administered following total thyroidectomy. It was only provided when blood test indicated hypocalcemia (serum total calcium concentration lower than 8.5 mg/dL) or when signs and symptoms of hypocalcemia, such as tingling or muscle cramps were reported. If no complications developed, the patients were discharged and followed up in the outpatient department two weeks after surgery. During follow-up, patients were evaluated for improvement of compression symptoms, surgical site healing, thyroid function, electrolyte conditions, and any delayed complications.

2.3. Clinical and Pathologic Characteristics

Clinical characteristics, including age, sex, body mass index (BMI), family history of thyroid disease, functional status, comorbidities, modified frailty index (mFI) score, history of other cancers, compression symptoms, preoperative image evaluation, ASA classification, size of the dominant tumor, fine-needle aspiration cytology (FNAC) results, use of IONM or energy devices, operation time, intraoperative blood loss, length of postoperative stay, in-hospital complications, postoperative survival time, and follow-up period, were collected.

The patients' preoperative functional status was classified into three categories, independent, partially dependent, and totally dependent, as assessed 30 days prior to the operation. Independence means that no assistance is required for daily activities, even with prosthetics or devices. Partial dependence refers to the need for assistance in daily activities, regardless of the use of equipment or aids [11]. Chronic kidney disease (CKD) was defined as an estimated glomerular filtration rate (eGFR) below 60 mL/min/1.73 m^2 and categorized by stage according to the KDIGO guideline. Compression symptoms, including palpable mass, foreign body sensation, difficulty in breathing, dysphagia, and voice change, were analyzed. The size of the dominant tumor was determined using neck ultrasonography or computed tomography (CT). The results of TI-RADS (Thyroid Imaging Reporting and Data System) were collected. FNAC results were based on the Bethesda System for Reporting Thyroid Cytopathology.

The duration of surgery was defined as the time from the initial skin incision to the completion of skin closure. The weight and volume of the excised thyroid glands were evaluated based on pathology reports. The events of postoperative intensive care unit (ICU) stay, delayed extubation, and blood transfusions were recorded. Postoperative complications, such as bleeding events, recurrent laryngeal nerve (RLN) injury, hypocalcemia, wound infection, pneumonia, cardiovascular events, and mortality, were recorded. Return to the emergency department and the need for readmission within 30 days of the procedure were recorded. The classifications of the thyroid pathologies were based on the final permanent pathological reports, which were reviewed by experienced pathologists. The benign thyroid pathology included nodular goiter, follicular adenoma, follicular neoplasm, and follicular nodular disease. The malignant tumors included papillary thyroid carcinoma, poorly differentiated carcinoma, Hürthle cell carcinoma, and follicular carcinoma.

This study was approved by the Institutional Review Board of Taichung Veterans General Hospital (TCVGH-IRB No. CE23063C, 7 March 2023) and ended on 30 June 2023.

2.4. Statistical Analysis

Statistical analyses were performed using SPSS ver. 22 (IBM Corp., Armonk, NY, USA). Chi-square or Fisher's exact tests were used to analyze categorical variables. For continuous variables, the Shapiro–Wilk test was used to test the normality of the sample distribution. A paired t-test or Mann–Whitney U test was used to compare continuous variables. Statistical significance was set at $p < 0.05$.

3. Results

3.1. Clinical Characteristics of the Patients before Thyroidectomy

In this retrospective cohort study, 17 patients aged 80 years and older were analyzed.

Benign disease was found in 12 of 17 (70.6%) patients, and malignancy was confirmed in five (29.4%) patients. Based on the pathological report, the patients were divided into two groups: benign and malignant. The clinical characteristics of each group are summarized in Table 1. The mean age was 85.6±4.8 years, with a notable female predominance (14 patients, 82.4%). Patients with benign diseases exhibited a mean age of 84.4 ± 2.9 years, which was lower than that of patients with malignant diseases (88.4 ± 7.4 years). However, this difference was not significant ($p = 0.339$).

The functional status of most patients was independent (10 patients, 58.8%), followed by partially dependent (7 patients, 41.2%). None of the patients were completely dependent.

The participants exhibited chronic comorbid conditions, including hypertension in 12 patients (70.6%) and CKD in 9 patients (52.9%, Supplementary Table S1). Chronic anemia, type 2 diabetes mellitus (DM), history of other cancers, and chronic pulmonary disease were observed in 35.3%, 30.4%, 17.6%, and 11.8% of patients, respectively. One patient had end-stage renal disease and underwent hemodialysis.

Table 1. Preoperative clinical characteristics of patients aged 80 years and older.

Characteristics	Overall (N = 17)	Benign (N = 12)	Malignant (N = 5)	*p*-Value
Age [†] (years)	85.6 ± 4.8	84.4 ± 2.9	88.4 ± 7.4	0.339
Sex, female, n (%)	14 (82.4%)	10 (83.3%)	4 (80%)	1.000
BMI [†]	24.5 ± 4.3	24.2 ± 4.9	25.1 ± 2.4	0.383
Family history of thyroid disease, n (%)	1 (6%)	0	1 (20%)	0.294
History of thyroidectomy	2 (11.8%)	1 (8.3%)	1 (20%)	0.515
Functional status, n (%)				1.000
Independent	10 (58.8%)	7 (58.3%)	3 (60%)	
Partially dependent	7 (41.2%)	5 (41.7%)	2 (40%)	
Totally dependent	0	0	0	
Comorbidity, n (%)				
Type 2 Diabetes mellitus	5 (30.4%)	2 (16.7%)	3 (60%)	0.117
Hypertension	12 (70.6%)	8 (66.7%)	4 (80%)	1.000
Cardiac vascular disease	2 (11.8%)	1 (8.3%)	1 (20%)	0.515
Chronic kidney disease	9 (52.9%)	7 (58.3%)	2 (40%)	0.620
Chronic pulmonary disease	2 (11.8%)	2 (16.7%)	0	1.000
Chronic anemia	6 (35.3%)	3 (25%)	3 (60%)	0.280
Other cancers	3 (17.6%)	2 (16.7%)	1 (20%)	1.000
Modified Frailty Index [†,††]	0.16 ± 0.12 (0.2 [0–0.4])	0.15 ± 0.12 (0.15 [0–0.4])	0.18 ± 0.11 (0.2 [0–0.3])	0.574
All compression symptoms, n (%)	14 (82.4%)	10(83.3%)	4 (80%)	1
Palpable mass or foreign body sensation	11 (64.7%)	7 (58.3)	4 (80%)	0.600
Difficulty in breathing	9 (52.9%)	6 (50%)	3 (60%)	1.000
Dysphagia	6(35.3%)	4 (33.3%)	2 (40%)	1.000
Voice change	2 (11.8%)	1(9.3%)	1 (20%)	0.515
Preoperative evaluation, n (%)				
Neck CT scan	10 (58.8%)	6 (50%)	4 (80%)	0.338
TI-RADS T1/T2/T3/T4/T5, n	0/8/4/5/0	0/7/1/4/0	0/1/3/1/0	0.667
Echocardiography	11 (64.7%)	7 (58.3)	4 (80%)	0.600
Lung function test	5 (29.4%)	3 (25%)	2 (40%)	0.472
ASA-1/2/3/4, n	0/6/11/0	0/4/8/0	0/2/3/0	0.605
Dominant tumor size [†] (cm)	5.5 ± 2.8	5.5 ± 2.3	5.6 ± 4.2	1
FNAC	15 (88.2%)	11 (91.7%)	4 (80%)	0.447
Time [††]	1 [0–17]	1 [0–17]	2 [0–9]	1.000
Unilateral %	10 (66.7%)	8 (72.7%)	3 (60%)	0.407
Bilateral %	5 (33.3%)	3 (27.2%)	2 (40%)	0.407
Bethesda category				
Nondiagnostic	4 (26.7%)	2 (16.7%)	2 (40%)	
Benign	8 (53.3%)	8 (66.7%)	0	
Atypia	2 (13.3%)	1 (8.3%)	1 (20%)	
Suspicious for malignancy	0	0	0	
Malignancy	1 (6.7%)	0	1 (20%)	

Data presented as [†] Mean ± SD, [††] Median (range). CT—computed tomography; FNAC—fine-needle aspiration cytology.

Fourteen (82.4%) patients exhibited compression-related symptoms. Palpable masses or foreign body sensations (64.7%) and difficulty in breathing (52.9%) were the most frequently reported. Two patients (11.8%) experienced significant voice changes.

The TI-RADS classification results and percentage of malignant pathology for the study population indicated that one out of eight patients (12.5%) in the TR2 group, three out of four patients (75%) in the TR3 group, and one out of five patients (20%) in the TR4 group had a malignant result, respectively.

More than half of the patients underwent neck CT (58.8%) and cardiac echocardiography (64.7%) for preoperative evaluation. Five (29.4 %) patients underwent lung function tests. Eleven patients were classified as ASA grade III (64.7%), and six patients were classified as ASA grade II (35.3%). The median mFI score of our patients was 0.2 (range, 0–0.4) and showed no difference between both groups.

Fifteen (88.2 %) patients underwent FNAC for preoperative diagnosis. Ten patients (66.7%) underwent unilateral aspiration, and the remaining underwent bilateral aspiration.

FNAC was performed on 59 nodules across all patients. One (1.7%) major complication was severe hematoma with airway obstruction.

Based on the preoperative FNAC examinations using the Bethesda System for Reporting Thyroid Cytopathology, the incidence rates of non-diagnostic, benign, atypical, and malignant tumors were 26.7%, 53.3%, 13.3%, and 6.7%, respectively. The mean size of the dominant tumor was 5.5 ± 2.8 cm. The largest tumor was 10.6 cm.

3.2. The Surgical Outcome of the Patients after Thyroidectomy

The surgical outcomes are summarized in Table 2. Unilateral total thyroidectomy was performed in 13 patients (76.4%); the remaining patients underwent bilateral total thyroidectomy (23.6%). IONM and energy devices were applied in 12 patients (70.6%).

The mean operation time for the benign group was 164.3 ± 32.0 min, which was notably shorter than the 231.0 ± 79.1 min observed in the malignant group ($p = 0.048$). The mean estimated blood loss for the benign group was 33 ± 75 mL, which is lower than the 192 ± 189 mL observed in the malignancy group. However, this difference was not significant ($p = 0.12$). The median length of postoperative stay was 2 days (range, 1–15 days).

No postoperative bleeding events, RLN injuries, wound infections, or in-hospital mortality occurred in this cohort study. Two of the seventeen patients (11.8%) experienced transient hypocalcemia. None of the patients had surgery-related pulmonary complications, such as pneumonia or respiratory failure. Two patients required ICU stay, delayed extubation, and blood transfusion (11.8%). The reason for ICU admission was unrelated to thyroid surgery. One patient developed hematoma and respiratory failure after a FNAC test, which necessitated ICU admission before surgery. The other patient had a large ulcerative thyroid tumor with neck cellulitis and underwent total thyroidectomy with a left pectoralis major myocutaneous (PMMC) flap for wound reconstruction. For safety considerations, postoperative ICU admission, blood transfusion, and monitoring the condition of the flap were necessary.

One patient (5.9%) visited the emergency room within 30 days of the operation, which was attributed to the acute exacerbation of chronic obstructive pulmonary disease caused by pneumonia. The other patients were not readmitted postoperatively. One patient with dementia visited the emergency room 60 days after surgery due to poor medication compliance, which led to hypothyroidism.

Until the end of the study in June 2023, the median postoperative follow-up duration was 28 months (range: 2–91 months). Thirteen (76.5%) patients were alive at the end of the study period. Two patients died of pneumonia (not surgery-related) at 18 and 46 months postoperatively; one patient died from newly diagnosed adenocarcinoma of an unknown primary site with liver and lung metastases, which was diagnosed 1 month postoperatively and died due to rapidly progressive disease 2 months postoperatively; one patient died from frailty 9 months postoperatively. These four patients had a median follow-up duration of 13.5 months (range, 2–46 months).

The clinicopathological results of five patients in the malignant group are summarized in Tables 3 and 4.

3.3. Two Reports of Rare Presentation

Two patients aged 98 and 92 years had rare presentations before thyroidectomy.

3.3.1. Case 1

The oldest patient was a 98-year-old woman with a 15-year history of hyperthyroidism, managed with carbimazole and bisoprolol. She had underlying hypertension, type 2 diabetes mellitus, and chronic kidney disease and was totally independent, with her daily living activities only assisted by a walker. The initial symptoms of neck swelling and palpitations appeared when she was in her 80s. Since then, she has received medical treatment for hyperthyroidism and multinodular goiter. Surgical intervention was suggested; however, her family refused due to her old age.

Table 2. Surgical outcomes of patients aged 80 years and older.

Characteristics	Overall (N = 17)	Benign (N = 12)	Malignant (N = 5)	*p*-Value
Surgery				0.538
Unilateral thyroidectomy, n (%)	13 (76.4%)	10 (83.3%)	3 (60%)	
Bilateral thyroidectomy, n (%)	4 (23.6%)	2 (16.7%)	2 (40%)	
IONM, n (%)	12 (70.6%)	8 (66.7%)	4 (80%)	1.000
Energy devices, n (%)	12 (70.6%)	8 (66.7%)	4 (80%)	1.000
Operation time †,†† (min)	183.9 ± 57.0 (180 [130–350])	164.3 ± 32.0 (157.5 [130–230])	231.0 ± 79.1(200 [155–350])	0.048
Unilateral thyroidectomy (min)	172.0 ± 43.3 (155 [130–270])	161.1 ± 34.4 (145 [130–230])	208.3 ± 58.0(200 [155–270])	0.143
Bilateral thyroidectomy (min)	222.5 ± 85 (222 [180–350])	180.0 ± 0(180 [180–180])	265.0 ± 120.2(265 [180–350])	1.000
Blood loss †,†† (mL)	80 ± 135 (0 [0–400])	33 ± 75 (0 [0–250])	192 ± 189 (210 [0–400])	0.12
Thyroid volume *,†,†† (mL)	97.9 ± 86.1 (86.2 [5.5–284.5]	91.9 ± 82.6 (71.9 [14.7–284.5])	112.5 ± 102.7 (103.5 [5.5–227.8])	0.861
Each Thyroid volume ***,†,†† (mL)	97.9 ± 86.1 (60.4 [5.5–284.5]	78.7 ± 75.2 (58.9 [14.7–284.5])	69.1 ± 52 (103 [5.5–113.9])	0.820
Unilateral thyroidectomy †,†† (mL)	70.3 ± 76.4 (57.5 [5.5–284.5])	78.6 ± 82.6 (57.5 [14.7–284.5])	42.9 ± 53 (19.6 [5.5–103.5])	0.469
Bilateral thyroidectomy †,†† (mL)	187.7 ± 46.6 (201.2 [120.7–227.8])	158.5 ± 53.5 (158.5 [120.7–196.4])	216.9 ± 15.4 (216.9 [206–227.8])	0.333
Thyroid weight †,††,** (g)	117.3 ± 131.2 (90 [10–510])	96.0 ± 84.6 (90 [10–300])	160.0 ± 201.7 (110 [10–510])	0.743
Each thyroid weight **** (g)	88.3 ± 85.6 (65 [10–300])	84.5 ± 85.2 ([10–300])	69.1 ± 52 (65 [10–255]	0.979
Unilateral thyroidectomy †,†† (g)	80.9 ± 84.8 (70 [10–300])	91.3 ± 95.2 (80 [10–300])	53.3 ± 51.3 (40 [10–110])	0.673
Bilateral thyroidectomy †,†† (g)	217.5 ± 195.2 (125 [110–510])	115.0 ± 7.1 (115 [110–120])	320.0 ± 268.7 (320 [130–510])	0.333
Length of stay †,†† (days)	3.1 ± 4.2 (2 [1–15])	1.7 ± 0.8 (1.5 [1–3])	6.6 ± 6.8 (2 [1–15])	0.134
Time to operation †,†† (months)	98.9 ± 143.3 (12 [1–500])	114.6 ± 162.5 (42 [2–500])	61.4 ± 83.6 (3 [1–180])	0.737
Postoperative follow-up time †,†† (months)	33.2 ± 27.6 (28 [2–91])	32 ± 28 (24.5 [2–91])	36 ± 29.5 (28 [9–85])	0.646

Data presented as † Mean ± SD, †† Median (range). * Total thyroid volume was calculated as length × width × height × 0.479. ** Two patient's thyroid weight data were lost. *** Each thyroid volume was calculated as length × width × height × 0.479/ǝ; Unilateral thyroidectomy ǝ= 1, Bilateral thyroidectomy ǝ= 2. **** each thyroid weight was calculated as thyroid weight/ǝ. IONM—intraoperative neurophysiological monitoring.

Table 3. Clinical information of the five patients with malignancy.

Patient No.	Age	Sex	Functional Status	Comorbidity	mFI	Indication for Surgery	Patient Referred from Specialist	FNAC	Operation	Operation Time (min)	Postoperative Complica-tion	Postoperative Stay (Days)	Follow-Up Time (Months)
1	98	F	independent	DM, HTN, CKD, Hyperthyroidism	0.2	Hematoma after FNAC caused respiratory failure	Endocrinologists	Non-diagnostic	Left thy-roidectomy	270	nil	15 *	19 ª
2	92	F	dependent	HTN	0.2	Rapid progression of neck mass with skin ulceration	Endocrinologists	Squamous cell carcinoma	Total thy-roidectomy + PMMC flap	350	Transient hypocalcemia	13 *	28 ª
3	90	M	independent	DM, HTN, CKD, Urothelial carcinoma	0.2	Difficulty in breathing and mass sensation	Urologist	Benign	Total thy-roidectomy	180	nil	2	9 *
4	82	F	independent	nil	0	Multiple pulmonary nodules post wedge resection with pathologic confirmation of metastatic thyroid carcinoma.	Thoracic surgeon	nil	Right thy-roidectomy ℮	155	Transient hypocalcemia	1	85 ª
5	80	F	dependent	DM, HTN, Hyperthyroidism, Parkinsonism, atrial fibrillation	0.3	Chest tightness, difficulty in breathing	Endocrinologists	atypia cell	Left thy-roidectomy ℮	200	nil	2	39 ª

* ICU admission; * death due to frailty; ª remained alive until the end of the study; ℮ history of thyroidectomy. CKD—chronic kidney disease; HTN—hypertension; DM—diabetes; FNAC—fine-needle aspiration cytology; PMMC—pectoralis major myocutaneous.

Table 4. Pathological results of the five patients with malignancy.

Patient No.	Age	Sex	Pathology	Associated Benign Thyroid Disease in Pathology	Dominant Tumor in Image Size (cm)	Pathological Cancer Size (cm)	Thyroid Weight (g)	Thyroid Volume(mL)	Multifocal	Angioinvasion	Lymphatic Invasion	Perineural Invasion	Extrathyroidal Extension
1	98	F	papillary microcarcinoma	Nodular goiter	4.6	0.2	110	103.5	Unifocal	nil	nil	nil	nil
2	92	F	poorly to undifferentiated carcinoma with prominent squamous differentiation, bilateral	Follicular neoplasm of uncertain malignant potential, left	10.6	8	510	227.8	Unifocal	nil	nil	nil	Present
3	90	M	Hürthle cell carcinoma, left	Nodular goiter, right	9.3	9	130	206.0	Unifocal	nil	nil	nil	nil
4	82	F	Papillary carcinoma, follicular variant, right	nil	0.5	0.5	10	5.5	Unifocal	nil	nil	nil	nil
5	80	F	Follicular carcinoma, encapsulated angioinvasive, with focal poorly differentiated thyroid carcinoma (5%)	nil	3.9	3.2	40	19.6	Unifocal	Present	nil	nil	nil

Six months prior to surgery, the patient experienced progressive worsening of symptoms, including dysphagia, odynophagia, and hoarseness. Thyroid ultrasonography revealed multiple bilateral nodules and mixed solid cystic calcifications. An endocrinologist then performed FNAC on 26 October 2021. Subsequently, the patient experienced neck swelling, pain, and stiffness after 5 min. Follow-up sonography revealed a hematoma at the puncture site without active bleeding. Compression was applied, and monitoring in the emergency department was recommended. However, the patient went home after 40 min of compression.

Later that day, she experienced difficulty in breathing and loss of consciousness at home. Cardiac massage was briefly performed by her son, after which her consciousness was regained. Upon arrival at the nearest emergency department, her Glasgow Coma Scale score was E4V1M1; endotracheal intubation was performed because of airway compression. Subsequently, the patient was transferred to the ICU of a local hospital.

The neck CT scan on the day of admission revealed bilaterally enlarged thyroid glands (Figure 1A). The right-side thyroid measured 4.8 × 3.4 × 6.7 cm, and the left-side thyroid measured 3.9 × 4.6 × 9.1 cm, in addition to multiple nodules, intrathoracic extension, active bleeding, and a hematoma measuring 5 × 2.3 × 4.1 cm in the left neck. Surgical intervention was not recommended by the surgeon, and tracheostomy was suggested because of the difficulty in weaning from ventilator support. However, her family opted to transfer her back to our hospital on 11 November 2021 for surgical evaluation.

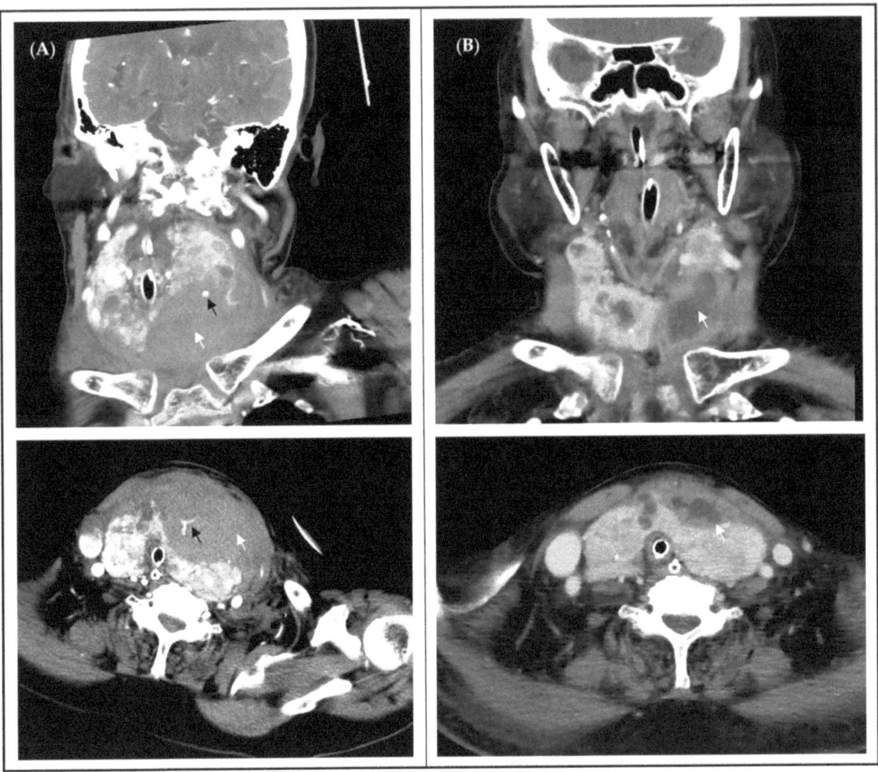

Figure 1. A 98-year-old woman developed hematoma after FNAC. (**A**): The neck CT showed enlargement of bilateral thyroid glands with multiple nodules, active bleeding (black arrow), and hematoma (white arrow), causing upper airway and left jugular vein compression with tracheal deviation. (**B**): The follow-up CT scan two weeks after FNAC showed hematoma (white arrow) in regression.

A subsequent neck CT scan performed at our hospital revealed a regressing hematoma (Figure 1B). Thirty-four days after fine-needle aspiration, left thyroidectomy was performed for airway decompression on 19 November 2021, and the patient remained in the ICU for postoperative care. The endotracheal tube was removed on postoperative day 5. Fifteen days postoperatively, the patient was discharged in a stable condition. Pathological examination revealed nodular goiter and papillary microcarcinoma. Follow-up in the outpatient department confirmed that the patient had maintained euthyroid status.

3.3.2. Case 2

A 92-year-old woman with a decades-long history of goiter was initially diagnosed at another hospital; the exact FNAC report remains uncertain owing to a lack of available records. Surgical intervention was proposed as a management strategy, but the patient declined it. Since then, the patient has been lost to follow-up.

Before the operation, she experienced progressive enlargement of the neck mass and a painful sensation associated with skin ulceration for 1 month. The neck CT scan performed on 2 February 2021 revealed a thyroid lesion measuring 8.8 × 9.1 × 10.6 cm with skin ulceration (Figure 2). The patient was admitted for the management of cellulitis and the neck mass.

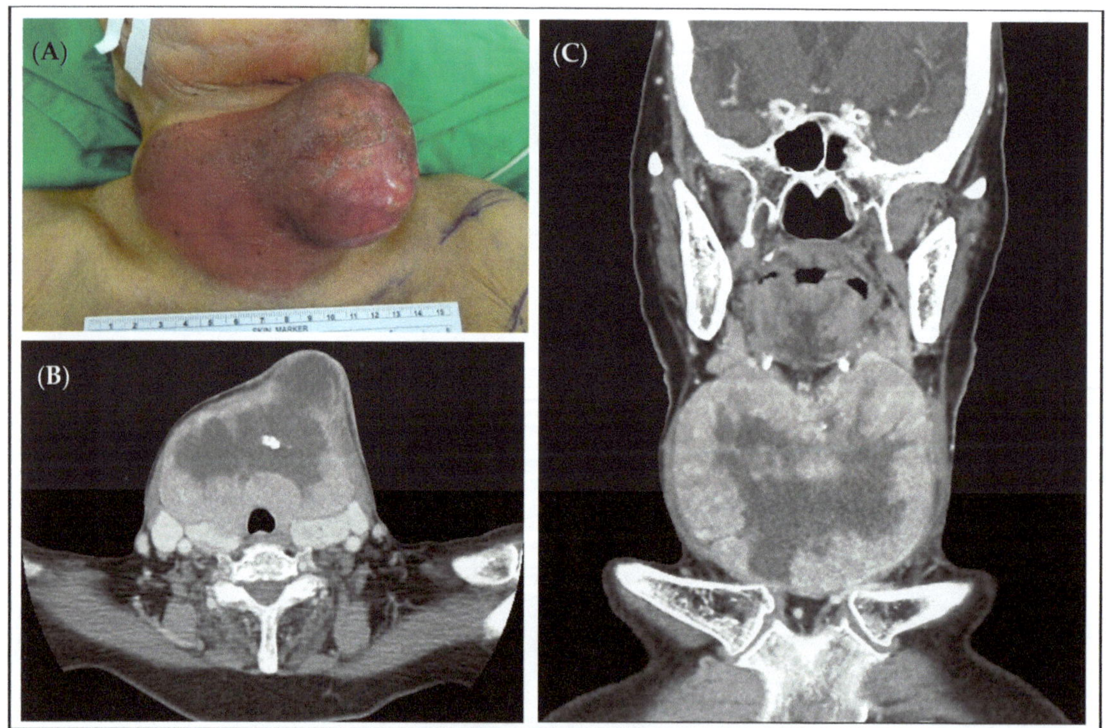

Figure 2. (A): A 92-year-old woman suffered from a large thyroid tumor with skin ulceration and cellulitis. **(B,C)**: The neck CT scan showed an enhancing necrotic mass 10 × 11 × 10 cm with calcification in lower neck submental space, abutting bilateral thyroid glands.

The FNAC test of the neck mass revealed thyroid squamous cell carcinoma (SCC). Considering the significant size of the tumor and skin involvement, total thyroidectomy with a left PMMC flap for wound reconstruction was performed on 27 February 2021. (Figure 3).

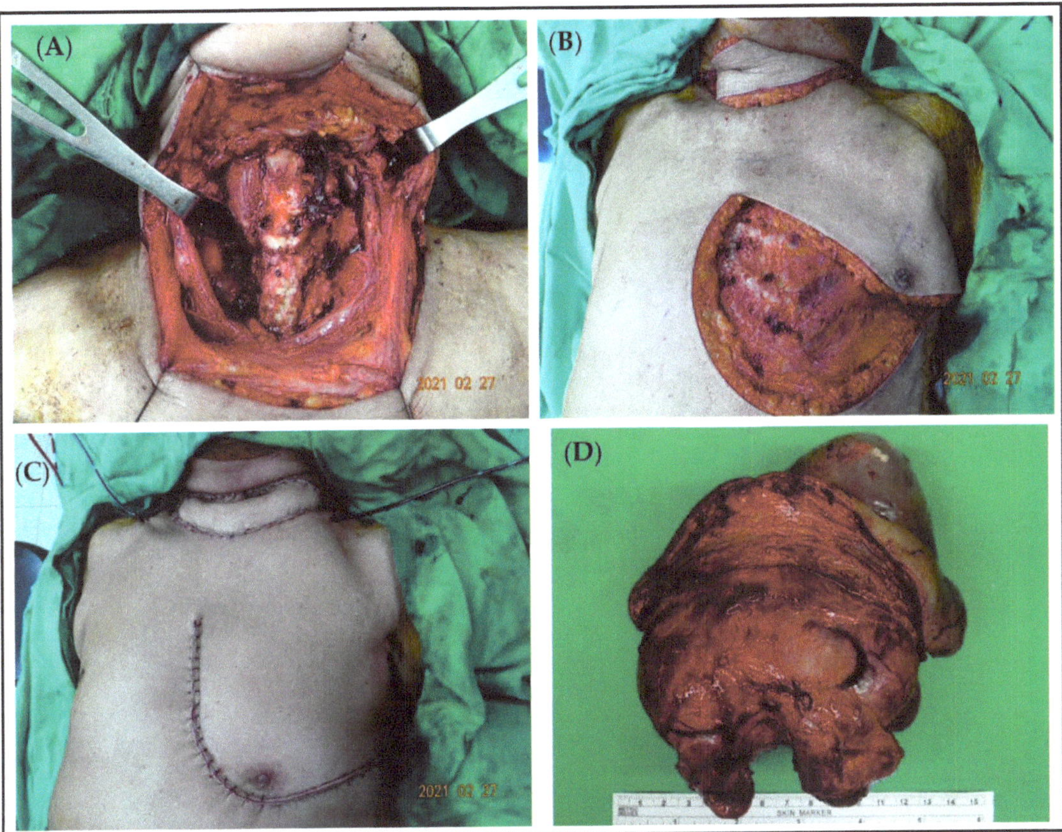

Figure 3. (**A**): The wound of a 92-year-old woman with a large thyroid tumor after total thyroidectomy. (**B,C**): The wound before and after PMMC flap for wound reconstruction. (**D**): A soft tumor, $8 \times 5 \times 3.2$ cm, with invasion into bilateral lobes of thyroid and skin tissue.

Postoperatively, the patient was transferred to the ICU care unit to monitor the flap condition. Extubation was delayed until postoperative day 5 because of the neck wound swelling, which may cause respiratory failure. Thirteen days postoperatively, the patient was discharged in a stable condition. The final pathological result was poor-to-undifferentiated carcinoma with prominent squamous differentiation.

4. Discussion

In this retrospective study, the overall patient cohort comprised 82.4% women, with a malignancy rate of 29.4%, consistent with the findings of several large-scale analyses [7,8,12]. The mean age of the patients was 85.6 years, which is higher than the average life expectancy in Taiwan (80.86 years) [13].

As the population continues to age, the frequency of surgical procedures involving older individuals is expected to increase. The diagnosis and treatment protocols for thyroid nodules are not influenced by age; therefore, the number of older patients requiring these surgical procedures is likely to increase.

Compression-related symptoms were the most common indications for thyroidectomy in this study. Fourteen (82.4%) patients underwent thyroidectomy to treat compression-related symptoms. Additionally, one patient underwent thyroidectomy due to post-FNAC complications, including hematoma and respiratory failure. Another study demonstrated that common surgical indications in older patients include hyperthyroidism resistant to medical management, symptoms of compression, and suspicion of malignant nodules [14]. Four of the patients underwent bilateral thyroidectomy, and the other ten underwent unilateral thyroidectomy. David et al. demonstrated that unilateral and bilateral thyroidectomy are equally effective for the relief of compressive symptoms [15]. Shared decision-making allows patients to choose the surgical procedure after being informed of the risks and benefits associated with both.

The study shows a 26.7% rate of non-diagnostic FNAC results. Given the retrospective nature of this study, data limitations make it difficult to determine the exact reasons. However, possible causes could include insufficient sampling or challenges in sampling larger or more complex cystic tumors.

The median time from the diagnosis of a thyroid tumor to thyroidectomy was 12 months (range, 1 month to 40 years). Concerns regarding potential postoperative complications among older patients and their families have led to considerable delays in the decision to undergo surgery. This could be attributed to the high incidence of complications observed in patients aged over 65 years, as reported in previous studies [7,8].

The incidence of perioperative cardiac complications is high in older patients aged over 80 years with coronary artery disease [16]. Preoperative electrocardiogram and echocardiography may provide important information for patients at an increased risk of cardiac complications [17]. To date, our hospital has updated preoperative protocol, and all patients aged 80 years and older routinely undergo preoperative echocardiography.

The average duration of our thyroidectomy procedures was 183.9 ± 57.0 min. The malignancy group (231.0 ± 79.1 min) experienced significantly longer operation times compared to the benign group (164.3 ± 32.0 min) ($p = 0.048$). However, owing to the limited sample size, these differences did not achieve significance in the unilateral and bilateral thyroidectomy subgroup analyses ($p = 0.143$, $p = 1$). The increased operation time in the malignant group may be due to extensive tissue invasion requiring resection and wound reconstruction, premalignant desmoid reaction, tumor adherence to adjacent structures with complicated dissection, and enhanced vascularity demanding meticulous hemostasis. Another explanation for this difference is that more patients with benign conditions underwent unilateral thyroidectomy (83.3%) than those with malignant conditions (60%), although this difference was not significant ($p = 0.163$).

In our study, the size of the dominant thyroid tumor was 5.5 ± 2.8 cm. Dellal et al. compared thyroidectomy in patients aged ≥ 65 and <65 years old, revealing that the mean nodule diameter was significantly larger in geriatric patients [18]. Based on our clinical experience, the operative time of older patients was longer than those of younger patients, possibly due to larger tumor size. Further study is needed to examine the differences and associations between tumor size and operative times in older and younger patients (Supplementary Table S2).

A majority of post-thyroidectomy complications in our patient groups were transient hypocalcemia, which occurred in two patients (11.8%). Some older patients are less sensitive to the symptoms of hypocalcemia and are less likely to report symptoms. Routine monitoring of serum calcium levels within 12–24 h post thyroidectomy may alleviate hypocalcemia symptoms and enhance the overall safety of patients [19]. We have adapted to this protocol. In our study, no major complications such as bleeding, RLN injury, wound infections, pneumonia, cardiovascular events, in-hospital mortality, or instances of readmission and reoperation were reported. Postoperative complication rates in previous studies varied from 1.2% to 6.8% [7–10,12,20]. Moreover, recent research has suggested that the incidence of complications may increase among older patients [7,8]. Post-thyroidectomy hypocalcemia ranges from 6% to 16.2% [10,13]. In a large national dataset analysis, Sahli et al. showed that the overall complication rate after thyroidectomies was 1.3%, the need for reoperation was 0.8%, and the readmission rate was 2.3% [7]. Older age was associated with an increased overall risk of complications (OR = 2.67). This reduced major complication rate could be ascribed to the expertise of high-volume surgeons. Supporting this finding, other studies have indicated a notable decrease in both thyroid cancer recurrence rates and postoperative complications among patients treated by high-volume surgeons, underscoring the importance of surgical experience in improving patient outcomes [21,22].

Matched cohort studies conducted by Papoian et al. showed that a higher Charlson Comorbidity Index is associated with an elevated risk of postoperative complications [8]. However, calculating the Charlson Comorbidity Index is relatively complex. Frailty is associated with a higher incidence of complications and mortality [23,24]. Moreover, mFI is strongly correlated with increased rates of postoperative complications, readmission, reoperation, discharge to skilled care, longer hospital stays, and higher mortality rates. A meta-analysis conducted by Panayi et al. revealed that postoperative mortality was more prevalent among patients with frailty and an mFI score of 0.36 or higher compared to those with lower scores [24]. The mFI score is simpler to calculate using patient characteristics. In our retrospective statistical analysis, the median mFI score of the patients was 0.2 (range of 0–0.4). Careful selection of older patients for surgery can decrease postoperative complications.

One patient developed a hematoma after the FNAC test, which necessitated intubation for airway maintenance and ICU admission before surgery. The FNAC test is currently the simplest, safest, and most cost-effective method for the identification of malignant thyroid nodules. However, this may cause a hematoma and require further surgical treatment. In this study, a FNAC test was performed for 17 patients with 59 thyroid nodules. Major complications occurred in one patient (1.7%), which presented as a massive hematoma and caused airway obstruction. Cappelli et al. reported that the complication rate of fine needle aspiration was approximately 0.15%, and major complications were 0.075% [25]. In older patients with loose soft tissue, the use of antiplatelet medication increases the feeding vessels of tumors and may increase the risk of hematoma. Using color Doppler ultrasonography to detect hypervascularity in a thyroid tumor and scheduling the withholding of antiplatelet or anticoagulant medications may reduce the risk of post-procedural bleeding complications.

One patient with dementia visited the emergency department 2 months after undergoing total thyroidectomy due to the discontinuation of medication-related hypothyroidism. The age of older patients was associated with misunderstanding the medication and medical treatment, eventually decreasing compliance [26]. The prevalence of dementia also increases with age and worsens comprehension and compliance with medical treatment. Individual decision-making on prescription appropriateness in patients with dementia and closely monitoring their understanding of medication could enhance medication adherence and potentially reduce emergency department visits [27]. The case manager's diligent follow-up with the patient and reminders to caregivers could enhance medication compliance. Unilateral thyroidectomy in selected patients with thyroid cancer may prevent hypothyroidism and improve the quality of life of older patients [28]. Shared decision-making also plays a crucial role in the choice between unilateral or total thyroidectomy, not only offering benefits in the quality of life and oncological outcomes but also alleviating the burden on patients and caregivers.

Although our study provides an in-depth analysis of outcomes in patients aged 80 years and older, it has certain limitations. The results were based on the experience of a single center with only 17 patients. While we recorded the presence of chronic diseases, we did not collect detailed data on preoperative blood pressure levels or control status, which may influence perioperative and postoperative outcomes. Future studies that incorporate more comprehensive data collection in these areas would help clarify their impact on patient outcomes. The feedback on quality of life we received from patients and their families was valuable; it was inherently subjective and based on observational data rather than quantitative measures. Future studies using multicenter or national data banks would provide stronger and more complete evidence.

5. Conclusions

Thyroidectomy is feasible for carefully selected patients aged 80 years and older. It provides benefits not only in terms of oncological curative treatment but also in improving the quality of life, such as compression symptoms and wound condition.

Supplementary Materials: The following supporting information can be downloaded at: https://www.mdpi.com/article/10.3390/medicina60091383/s1, Table S1: Renal Function Assessment in patients aged 80 years and older; Table S2: Correlation Analysis Between Operation Time and Various Factors.

Author Contributions: Conceptualization: W.-H.C.; Methodology: W.-H.C. and Y.-J.C.; Validation: W.-H.C.; Analysis: W.H.; Writing—original draft preparation: W.H.; Writing—review and editing: W.-H.C. and W.H.; Visualization: W.H. All authors have read and agreed to the published version of the manuscript.

Funding: This research received no external funding.

Institutional Review Board Statement: The study was conducted in accordance with the Declaration of Helsinki and approved by the Institutional Review Board of Taichung Veterans General Hospital (protocol code TCVGH-IRB No. CE23063C, date of approval: 7 March 2023).

Informed Consent Statement: Informed consent was obtained from all subjects involved in the study. No identifiable information is present in the manuscript.

Data Availability Statement: The data presented in this study are available upon reasonable request from the corresponding authors (W.H. or W.-H.C).

Acknowledgments: During the preparation of this work the authors used ChatGPT to check grammar and semantics. After using this tool, the authors reviewed and edited the content as needed and take full responsibility for the content of the publication.

Conflicts of Interest: The authors declare no conflicts of interest. The funders had no role in the design of the study; in the collection, analyses, or interpretation of data; in the writing of the manuscript; or in the decision to publish the results.

References

1. National Development Council. *Population Projection Report of the Republic of China (Taiwan)*; National Development Council: Taipei, Taiwan, 2022; ISBN 978-626-7162-20-0.
2. World Population Prospects 2019, Highlights. Available online: https://www.un.org/development/desa/pd/news/world-population-prospects-2019-0 (accessed on 19 July 2024).
3. Ospina, N.S.; Papaleontiou, M. Thyroid Nodule Evaluation and Management in Older Adults: A Review of Practical Considerations for Clinical Endocrinologists. *Endocr. Pract.* **2021**, *27*, 261–268. [CrossRef]
4. Lim, H.; Devesa, S.S.; Sosa, J.A.; Check, D.; Kitahara, C.M. Trends in Thyroid Cancer Incidence and Mortality in the United States, 1974–2013. *JAMA* **2017**, *317*, 1338–1348. [CrossRef] [PubMed]
5. Grubey, J.S.; Raji, Y.; Duke, W.S.; Terris, D.J. Outpatient thyroidectomy is safe in the elderly and super-elderly. *Laryngoscope* **2018**, *128*, 290–294. [CrossRef] [PubMed]
6. Sahli, Z.T.; Zhou, S.; Najjar, O.; Onasanya, O.; Segev, D.; Massie, A.; Zeiger, M.A.; Mathur, A. Octogenarians have worse clinical outcomes after thyroidectomy. *Am. J. Surg.* **2018**, *216*, 1171–1176. [CrossRef] [PubMed]
7. Sahli, Z.T.; Ansari, G.; Gurakar, M.; Canner, J.K.; Segev, D.; Zeiger, M.A.; Mathur, A. Thyroidectomy in older adults: An American College of Surgeons National Surgical Quality Improvement Program study of outcomes. *J. Surg. Res.* **2018**, *229*, 20–27. [CrossRef] [PubMed]
8. Papoian, V.; Marji, F.P.; Rosen, J.E.; Carroll, N.M.; Felger, E.A. Safety of Thyroid Surgery in the Elderly: A Propensity Score Matched Cohort Study. *J. Surg. Res.* **2019**, *242*, 239–243. [CrossRef] [PubMed]
9. Diaconescu, M.R.; Glod, M.; Costea, I. Clinical Features and Surgical Treatment of Thyroid Pathology in Patients Over 65 Years. *Chirurgia* **2016**, *111*, 120–125. [CrossRef] [PubMed]
10. Al-Qahtani, K.H.; Tunio, M.A.; Asiri, M.A.; Bayoumi, Y.; Balbaid, A.; Aljohani, N.J.; Fatani, H. Comparative clinicopathological and outcome analysis of differentiated thyroid cancer in Saudi patients aged below 60 years and above 60 years. *Clin. Interv. Aging* **2016**, *11*, 1169–1174. [CrossRef]
11. Zhang, L.M.; Hornor, M.A.; Robinson, T.; Rosenthal, R.A.; Ko, C.Y.; Russell, M.M. Evaluation of Postoperative Functional Health Status Decline Among Older Adults. *JAMA Surg.* **2020**, *155*, 950–958. [CrossRef]

12. Gut, L.; Bernet, S.; Huembelin, M.; Mueller, M.; Baechli, C.; Koch, D.; Nebiker, C.; Schuetz, P.; Mueller, B.; Christ, E.; et al. Sex-Specific Differences in Outcomes Following Thyroidectomy: A Population-Based Cohort Study. *Eur. Thyroid. J.* **2021**, *10*, 476–485. [CrossRef]

13. Abridged Life Table in Republic of China Area. 2021. Available online: https://www.lifetable.de/File/GetDocument/data/TWN/TWN000020212021CU1.pdf (accessed on 21 August 2024).

14. Gervasi, R.; Orlando, G.; Lerose, M.A.; Amato, B.; Docimo, G.; Zeppa, P.; Puzziello, A. Thyroid surgery in geriatric patients: A literature review. *BMC Surg.* **2012**, *12* (Suppl. S1), S16. [CrossRef] [PubMed]

15. Moffatt, D.C.; Tucker, J.; Goldenberg, D. Management of compression symptoms of thyroid goiters: Hemithyroidectomy is equally as successful as total thyroidectomy. *Am. J. Otolaryngol.* **2023**, *44*, 103676. [CrossRef] [PubMed]

16. Liu, Z.; Xu, G.; Xu, L.; Zhang, Y.; Huang, Y. Perioperative Cardiac Complications in Patients Over 80 Years of Age with Coronary Artery Disease Undergoing Noncardiac Surgery: The Incidence and Risk Factors. *Clin. Interv. Aging* **2020**, *15*, 1181–1191. [CrossRef]

17. Bolat, İ. Preoperative Right Ventricular Echocardiographic Parameters Predict Perioperative Cardiovascular Complications in Patients Undergoing Non-Cardiac Surgery. *Heart Lung Circ.* **2020**, *29*, 1146–1151. [CrossRef]

18. Dellal, F.D.; Özdemir, D.; Tam, A.A.; Baser, H.; Tatli Dogan, H.; Parlak, O.; Ersoy, R.; Cakir, B. Clinicopathological features of thyroid cancer in the elderly compared to younger counterparts: Single-center experience. *J. Endocrinol. Investig.* **2017**, *40*, 471–479. [CrossRef]

19. Păduraru, D.N.; Ion, D.; Carsote, M.; Andronic, O.; Bolocan, A. Post-thyroidectomy Hypocalcemia—Risk Factors and Management. *Chirurgia* **2019**, *114*, 564–570. [CrossRef] [PubMed]

20. Echanique, K.A.; Govindan, A.; Mohamed, O.M.; Sylvester, M.; Baredes, S.; Ying, M.Y.; Kalyoussef, E. Age-Related Trends of Patients Undergoing Thyroidectomy: Analysis of US Inpatient Data from 2005 to 2013. *Otolaryngol. Head Neck Surg.* **2019**, *160*, 457–464. [CrossRef]

21. Gorbea, E.; Goldrich, D.Y.; Agarwal, J.; Nayak, R.; Iloreta, A.M. The impact of surgeon volume on total thyroidectomy outcomes among otolaryngologists. *Am. J. Otolaryngol.* **2020**, *41*, 102726. [CrossRef]

22. Eskander, A.; Noel, C.W.; Griffiths, R.; Pasternak, J.D.; Higgins, K.; Urbach, D.; Goldstein, D.P.; Irish, J.C.; Fu, R. Surgeon Thyroidectomy Case Volume Impacts Disease-free Survival in the Management of Thyroid Cancer. *Laryngoscope* **2023**, *133* (Suppl. S4), S1–S15. [CrossRef]

23. Wahl, T.S.; Graham, L.A.; Hawn, M.T.; Richman, J.; Hollis, R.H.; Jones, C.E.; Copeland, L.A.; Burns, E.A.; Itani, K.M.; Morris, M.S. Association of the Modified Frailty Index With 30-Day Surgical Readmission. *JAMA Surg.* **2017**, *152*, 749–757. [CrossRef]

24. Panayi, A.C.; Orkaby, A.R.; Sakthivel, D.; Endo, Y.; Varon, D.; Roh, D.; Orgill, D.; Neppl, R.; Javedan, H.; Bhasin, S.; et al. Impact of frailty on outcomes in surgical patients: A systematic review and meta-analysis. *Am. J. Surg.* **2019**, *218*, 393–400. [CrossRef]

25. Cappelli, C.; Pirola, I.; Agosti, B.; Tironi, A.; Gandossi, E.; Incardona, P.; Marini, F.; Guerini, A.; Castellano, M. Complications after fine-needle aspiration cytology: A retrospective study of 7449 consecutive thyroid nodules. *Br. J. Oral Maxillofac. Surg.* **2017**, *55*, 266–269. [CrossRef] [PubMed]

26. Albrecht, J.S.; Gruber-Baldini, A.L.; Hirshon, J.M.; Brown, C.H.; Goldberg, R.; Rosenberg, J.H.; Comer, A.C.; Furuno, J.P. Hospital discharge instructions: Comprehension and compliance among older adults. *J. Gen. Intern. Med.* **2014**, *29*, 1491–1498. [CrossRef] [PubMed]

27. Renom-Guiteras, A. Potentially inappropriate medication among people with dementia: Towards individualized decision-making. *Eur. Geriatr. Med.* **2021**, *12*, 569–575. [CrossRef]

28. Yaniv, D.; Vainer, I.; Amir, I.; Robenshtok, E.; Hirsch, D.; Watt, T.; Hilly, O.; Shkedy, Y.; Shpitzer, T.; Bachar, G.; et al. Quality of life following lobectomy versus total thyroidectomy is significantly related to hypothyroidism. *J. Surg. Oncol.* **2022**, *126*, 640–648. [CrossRef] [PubMed]

medicina

Single-Center Experience of Parathyroidectomy Using Intraoperative Parathyroid Hormone Monitoring

Seong Hoon Kim [†], Si Yeon Lee [†], Eun Ah Min, Young Mi Hwang, Yun Suk Choi [*,‡] and Jin Wook Yi [*,‡]

Department of Surgery, Inha University Hospital, College of Medicine, Incheon 22332, Korea
* Correspondence: yunsukki@gmail.com (Y.S.C.); jinwook.yi@inha.ac.kr (J.W.Y.); Tel.: +82-32-890-3437 (Y.S.C.); +82-32-890-3437 (Y.S.C.)
† These two authors contributed equally to first author in this article.
‡ These two authors contributed equally to the corresponding author in this article.

Abstract: *Background and Objectives*: Hyperparathyroidism (HPT) is a rare endocrine disease associated with the elevated metabolism of calcium, vitamin D, and phosphate by the hyperfunctioning of the parathyroid glands. Here, we report our experience of parathyroidectomy using intraoperative parathyroid hormone (IOPTH) monitoring in a single tertiary hospital. *Materials and Methods*: From October 2018 to January 2022, a total of 47 patients underwent parathyroidectomy for HPT. We classified the patients into two groups—primary HPT (PHPT, n = 37) and renal HPT (RHPT, n = 10)—and then reviewed the patients' data, including their general characteristics, laboratory results, and perioperative complications. *Results*: Thirty-five of the thirty-seven patients in the PHPT group underwent focused parathyroidectomy, while all ten patients in the RHPT group underwent subtotal parathyroidectomy. IOPTH monitoring based on the Milan criteria was used in all cases. Preoperative and 2-week, 6-month, and 12-month postoperative parathyroid hormone (PTH) levels were within the normal range in the PHPT group, whereas they were higher than normal in the RHPT group. Transient hypocalcemia occurred only in the RHPT group, with calcium levels returning to normal levels 12 months after surgery. *Conclusions*: Parathyroidectomy with IOPTH monitoring in our hospital showed favorable clinical outcomes. However, owing to the small number of patients due to the low frequency of parathyroid disease, long-term, prospective studies are needed in the future.

Keywords: parathyroid hormone; hyperparathyroidism; intraoperative parathyroid hormone

Citation: Kim, S.H.; Lee, S.Y.; Min, E.A.; Hwang, Y.M.; Choi, Y.S.; Yi, J.W. Single-Center Experience of Parathyroidectomy Using Intraoperative Parathyroid Hormone Monitoring. *Medicina* **2022**, *58*, 1464. https://doi.org/10.3390/medicina58101464

Academic Editor: Ioannis Ilias

Received: 26 September 2022
Accepted: 14 October 2022
Published: 16 October 2022

Publisher's Note: MDPI stays neutral with regard to jurisdictional claims in published maps and institutional affiliations.

1. Introduction

Hyperparathyroidism (HPT) is an endocrine disease related to the elevated metabolism of calcium (Ca), vitamin D, and phosphate (P) due to the excessive secretion of parathyroid hormone (PTH) [1–3]. HPT is classified into primary (PHPT), secondary (SHPT), and tertiary (THPT) disease depending on the disease mechanism. HPT raises serum Ca and P levels and lowers vitamin D levels, resulting in various diseases, including bone, kidney, gastrointestinal, neuropsychiatric, soft tissue, and cardiovascular disorders. A lack of proper treatment of the disease associated with HPT, especially cardiovascular complications, may lead to fatal consequences for the patient [1–3]. The treatment required for PHPT is surgical resection of the pathologic parathyroid gland; some medically intractable SHPT and THPT patients also require surgical treatment [4,5].

PHPT is a disease caused by the excessive production of PTH from one or more parathyroid glands. The incidence of PHPT is reported to be approximately 20–50 cases per 100,000 people annually, and it is the third-most common endocrine disorder in the United States [6,7]. Owing to the development of various diagnostic modalities, the number of cases of PHPT has been increasing [8]. Approximately 85% of PHPT cases result from a single parathyroid adenoma, while the remaining 15% of the cases are caused by multiple abnormal glands [9]. In the case of PHPT, surgical excision rather than pharmacological therapy is recommended as the gold standard treatment. Previously, surgery was performed to examine all parathyroid glands using a large incision, but focused parathyroidectomy using minimal skin incision has been introduced with the development of preoperative localizing imaging techniques and intraoperative PTH (IOPTH) monitoring to confirm the complete excision of the pathologic parathyroid glands [10–12].

SHPT is a condition characterized by elevated PTH secretion caused by abnormal vitamin D and calcium metabolism. The major cause of SHPT is chronic renal failure, but other causes include gastrointestinal absorption dysfunction, vitamin D deficiency, liver diseases, and pseudohypoparathyroidism [4,5]. THPT is a condition of persistent HPT after kidney transplantation [1,12]. SHPT and THPT are both caused by renal problems and can be classified as RHPT. In most cases of RHPT, pharmacological therapy using cinacalcet is recommended as the initial treatment. When this pharmacological therapy fails to produce a proper response, parathyroidectomy—total parathyroidectomy with autotransplantation or subtotal parathyroidectomy—should be recommended [13,14].

Because the incidence of HPT is very low, parathyroid surgery is performed in very few institutions in Korea [12,13,15–20]. There are even fewer centers that use the IOPTH assay for monitoring for parathyroid surgery in Korea [21–24]. A single endocrine surgeon at our hospital has been performing parathyroid surgery using IOPTH since 2018, and we report our initial results of parathyroid surgery in this paper.

2. Materials and Methods

2.1. Patients

From September 2018 to January 2022, a total of 47 patients received parathyroid surgery due to HPT. All patients who underwent parathyroidectomy at Inha University were included, and there were no special exclusion criteria. The clinical diagnoses were as follows: PHPT in 37 patients, SHPT in 9 patients, and THPT in 1 patient. In the analysis, the SHPT and THPT patients were included in the RHPT group as they shared the same origin of HPT (renal dysfunction). We retrospectively reviewed the electronic medical records for the patients' clinical information, laboratory test results (for Ca, ionized Ca, P, and PTH), surgical findings, hospitalization records, and pathologic reports.

2.2. Surgical Indication

In PHPT, surgery was performed on patients who had symptoms (nephrolithiasis, fractures, symptomatic hypercalcemia). For asymptomatic PTHP patients, surgery was performed if serum calcium was more than 1.0 mg/dL above the normal, creatinine clearance < 60 cc/min, nephrocalcinosis or nephrolithiasis identified on imaging, 24 h urine calcium > 400 mg/day, osteoporosis by bone density at any site (T score $\leq$ −2.5), clinical fragility fracture, vertebral compression fracture on spine imaging, or age < 50 years. If medical therapy was refractory, surgery was performed on SHPT patients with symptoms and markedly elevated PTH levels. That includes a persistent high serum level of intact PTH level > 500 pg/m, hyperphosphatemia (serum P > 6.0 mg/dL) and/or hypercalcemia (serum Ca > 10.0 mg/dL), deformity, fracture, progressive reduction in bone mineral content, ectopic calcification, or neuro-muscular psychiatric symptoms, etc. In THPT, surgery was performed on patients with persistent hypercalcemia more than 6 months after kidney transplant, low bone mineral density, renal stone or nephrocalcinosis, deterioration of kidney graft due to THPT, symptomatic HPT, or enlarged parathyroid gland detected by the US [2,3].

2.3. Surgery Methods and IOPTH Monitoring

All surgeries were performed by a single endocrine surgeon (JW Yi). A 2–3 cm skin incision was used to perform focused parathyroidectomy for PHPT. Before the surgery, the location of the pathologic parathyroid gland was determined by parathyroid SPECT-CT and ultrasound-guided skin marking. Subtotal parathyroidectomy (the removal of three and a half parathyroid glands) was applied for RHPT patients using a transverse skin incision of ~7–9 cm. Intraoperative neuromonitoring (NIM 3.0, Medtronic, Minneapolis, MN, USA) was used in all parathyroid surgeries.

The IOPTH assay was performed at four surgical time points: pre-incision, pre-excision, and 5 min and 10 min after the removal of the pathologic parathyroid gland. The Miami criterion (a >50% decrease in PTH 10 min after parathyroidectomy compared to the highest level of PTH before the excision) was indicated for the successful removal of the hyperfunctioning parathyroid gland in both focused and subtotal parathyroidectomy in our institution [25]. In focused parathyroidectomy, bilateral exploration was indicated when the PTH level was not successfully decreased. For the subtotal parathyroidectomy, all parathyroid glands were identified through bilateral exploration, saving half of the most normal-looking gland and removing the remaining three glands. Pre-excision PTH was sampled after one and a half glands were removed, while post-excision PTH was sampled 5 and 10 min after the complete removal of the three and a half glands. Laboratory tests for PTH, Ca, ionized Ca, and P were performed on the admission day, the day after surgery, and 6 and 12 months after surgery.

2.4. Statistics and Ethical Consideration

We used R programming language version 4.2.0 for the statistical analysis [15]. Chi-square or Fisher's exact test was applied to the cross-table analysis according to the sample size. The unpaired t-test was used for the mean comparison. Statistical significance was defined as a *p*-value < 0.05. Ethical approval for this study was obtained from the institutional review board of our hospital (IRB number: 2022-05-028).

3. Results

Table 1 presents the clinical and pathologic characteristics of patients. Among all 47 patients, 37 were female and 10 were male; the mean age was 56.50 ± 11.54 years. The clinical diagnoses were PHPT in 37 patients, SHPT in 9 patients, and THPT in 1 patient. The pathologic diagnoses were parathyroid adenoma and parathyroid hyperplasia in 31 and 16 patients, respectively. Among the 37 PHPT patients, one-gland parathyroidectomy was performed in 35 patients, two-gland parathyroidectomy was performed in 1 patient, and bilateral exploration with the removal of three glands was performed in 1 patient. Subtotal parathyroidectomy was performed in all 10 patients with RHPT. Although transient vocal cord palsy occurred in two patients, no permanent vocal cord palsy was observed. Hypertrophic scars were observed in two patients.

Table 2 presents the comparative results between PHPT and RHPT. Age did not differ between the two groups, but the proportion of female patients was significantly higher in the PHPT group than in the RHPT group (32/37 (86.5%) versus 5/10 (50.0%); *p* = 0.024). The major clinical manifestations of PHPT were ureteral or renal stones, osteoporosis, and a history of bone fracture. The operation time was longer (68.2 ± 42.8 min versus 83.0 ± 21.4 min, *p* = 0.139), the estimated blood loss was higher (10.8 ± 20.1 mL versus 130.0 ± 248.6 mL, *p* = 0.164), and the hospital stay days were longer (1.8 ± 1.1 days versus 6.0 ± 6.2 days, *p* = 0.059) in the RHPT group. In the PHPT group, the most common positions of the hyperfunctioning glands were the lower right (14 glands) and lower left (12 glands). The upper left was the most common location of the saved glands in the RHPT group.

Table 3 and Figures 1 and 2 show the laboratory findings of the PHPT and RHPT groups. The preoperative PTH level was significantly higher in the RHPT group (161.6 ± 95.4 pg/mL versus 1242.1 ± 1075.3 pg/mL; *p* = 0.011). The PTH level measured at 6 months and 12 months after surgery did not differ significantly between the two groups. According to the IOPTH assay results, the PTH levels at all time points were higher in the RHPT group than in the PHPT group. A >50% decrease in the PTH level was observed 10 min after the excision in all patients in the PHPT and RHPT groups. The preoperative Ca level was significantly higher in the PHPT group (10.9 ± 0.9 mg/dL versus 9.6 ± 1.7 mg/dL; *p* = 0.042). After surgery, the Ca level was normalized in the PHPT group, whereas hypocalcemia was observed in the RHPT group at 6 months after surgery (9.5 ± 1.3 mg/dL versus 6.8 ± 1.2 mg/dL; p = 0.003). No difference was observed in the Ca levels 12 months after surgery (8.6 ± 2.8 mg/dL versus 8.6 ± 2.07 mg/dL; p = 0.942). There was no significant difference in the ionized Ca level between the two groups before or after surgery. The P level was significantly higher in the RHPT group before surgery (2.6 ± 0.5 mg/dL versus 3.9 ± 1.1 mg/dL; p = 0.023) but decreased and increased again 12 months after surgery, without any significant difference between the PHPT and RHPT groups (3.3 ± 0.6 mg/dL versus 4.1 ± 1.3 mg/dL; p = 0.377).

Table 1. General characteristics of all hyperparathyroidism patients (n = 47).

Variables	Number of Patients
Age (years, mean ± sd)	56.5 ± 11.5
Sex	
Male	10
Female	37
Clinical diagnosis	
Primary hyperparathyroidism	37
Secondary hyperparathyroidism	9
Tertiary hyperparathyroidism	1
Parathyroidectomy extent	
One-gland parathyroidectomy	35
Two-gland parathyroidectomy	1
Bilateral exploration	1
Subtotal parathyroidectomy	10
Pathologic diagnosis	
Parathyroid adenoma	31
Parathyroid hyperplasia	16
Postoperative complications	
Transient vocal cord palsy	2
Permanent vocal cord palsy	0
Hypertrophic scar or keloid	2

Table 2. Comparison between primary and renal (secondary and tertiary) hyperparathyroidism.

Variables	PHPT (n = 37)	RHPT (n = 10)	p-Value
Age (years, mean ± sd)	57.1 ± 12.1	54.2 ± 9.3	0.425
Sex			
Male	5 (13.5%)	5 (50.0%)	0.024
Female	32 (86.5%)	5 (50.0%)	
BMI (kg/m², mean ± sd)	24.0 ± 4.0	24.0 ± 2.5	0.983
Comorbidity (number of patients)			
Diabetes	4	1	
Hypertension	9	7	
Chronic renal failure	0	10	
Coronary artery disease	2	0	
Arrhythmia	1	0	
Cerebrovascular disease	1	0	
Hepatitis B	1	0	
Tuberculosis	1	0	
Osteoporosis	4	0	
Fracture history	1	0	
Ureter or renal stone	7	0	
Operation time (min, mean ± sd)	68.2 ± 42.8	83.0 ± 21.4	0.139
Estimated blood loss (mL, mean ± sd)	10.8 ± 20.1	130.0 ± 248.6	0.164
Largest gland size (cm, mean ± sd)	1.8 ± 1.0	1.4 ± 1.1	0.407

Table 2. *Cont.*

Variables	PHPT (n = 37)	RHPT (n = 10)	*p*-Value
Hospital stay days after surgery (days, mean ± sd)	1.8 ± 1.1	6.0 ± 6.2	0.059
Pathologic gland location (Focused parathyroidectomy), number (%)			
Lower right	14 (36.8%)		
Upper right	5 (13.2%)		
Lower left	12 (31.6%)		
Upper left	7 (18.4%)		
Saved gland location (Subtotal parathyroidectomy), number (%)			
Lower right		1 (10.0%)	
Upper right		2 (20.0%)	
Lower left		1 (10.0%)	
Upper left		6 (60.0%)	

Table 3. Laboratory findings according to primary and renal (secondary and tertiary) hyperparathyroidism.

Variables	PHPT (n = 37)	RHPT (n = 10)	*p*-Value
PTH, preoperative (pg/mL)	161.6 ± 95.4	1242.1 ± 1075.3	0.011
PTH, pre-incision (pg/mL)	210.3 ± 168.5	1200.4 ± 773.4	0.005
PTH, pre-excision (pg/mL)	143.2 ± 90.8	415.0 ± 269.4	0.056
PTH, 5 min after excision (pg/mL)	64.0 ± 39.2	324.6 ± 258.2	0.087
PTH, 10 min after excision (pg/mL)	39.0 ± 25.0	171.7 ± 138.6	0.021
PTH, 6 months after surgery (pg/mL)	52.1 ± 31.8	71.9 ± 78.2	0.454
PTH, 12 months after surgery (pg/mL)	50.2 ± 27.0	108.7 ± 172.2	0.313
Calcium, preoperative (mg/dL)	10.9 ± 0.9	9.6 ± 1.7	0.042
Calcium, postoperative (mg/dL)	8.9 ± 0.8	7.9 ± 1.9	0.120
Calcium, 6 months after surgery (mg/dL)	9.5 ± 1.3	6.8 ± 1.2	0.003
Calcium, 12 months after surgery (mg/dL)	8.6 ± 2.8	8.6 ± 2.0	0.942
Ionized calcium, preoperative (mmol/L)	1.4 ± 0.1	1.2 ± 0.3	0.132
Ionized calcium, postoperative (mmol/L)	1.1 ± 0.1	1.0 ± 0.2	0.136
Ionized calcium, 6 months after surgery (mmol/L)	1.2 ± 0.1	1.1 ± 0.3	0.334
Ionized calcium, 12 months after surgery (mmol/L)	1.2 ± 0.2	1.1 ± 0.4	0.391
Phosphate, preoperative (mg/dL)	2.6 ± 0.5	3.9 ± 1.1	0.023
Phosphate, postoperative (mg/dL)	3.05 ± 0.9	3.8 ± 1.2	0.061
Phosphate, 6 months after surgery (mg/dL)	3.3 ± 0.7	2.7 ± 0.4	0.075
Phosphate, 12 months after surgery (mg/dL)	3.3 ± 0.6	4.1 ± 1.3	0.377

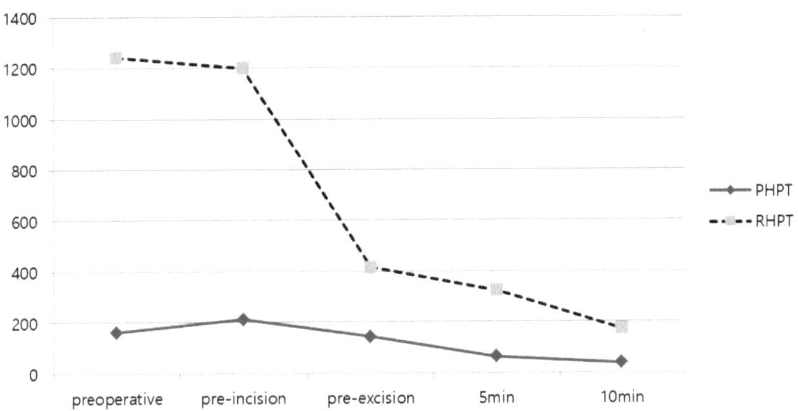

Figure 1. IOPTH assay according to primary hyperparathyroidism (PHPT) and renal (secondary and tertiary) hyperparathyroidism (RHPT).

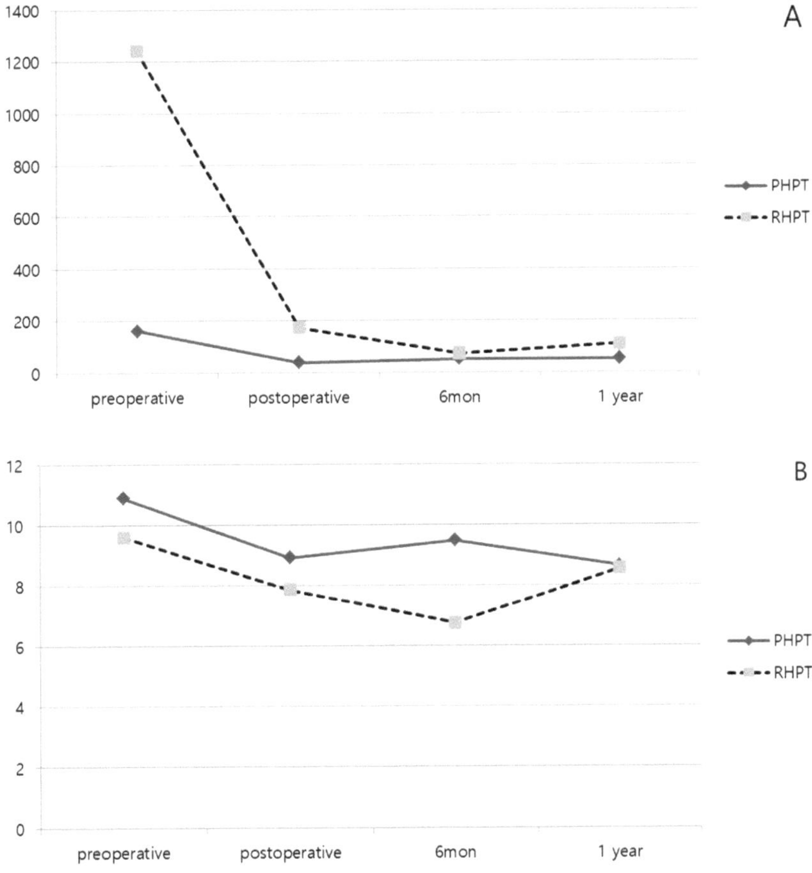

Figure 2. *Cont.*

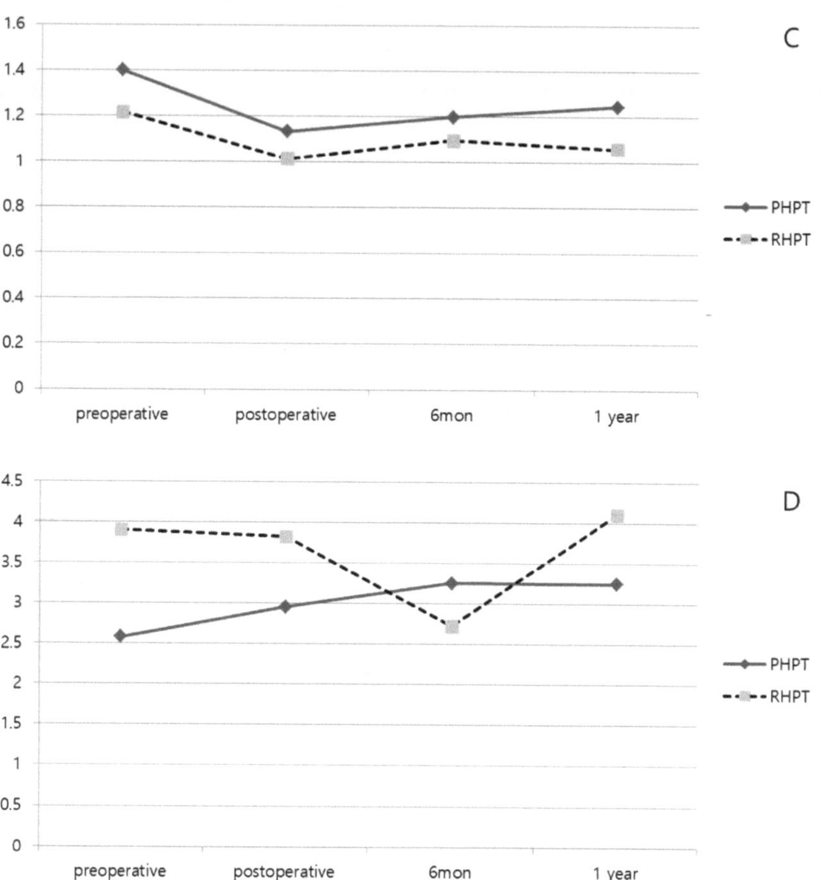

Figure 2. Laboratory results change before and after parathyroidectomy according to primary hyperparathyroidism (PHPT) and renal (secondary and tertiary) hyperparathyroidism (RHPT). ((**A**) PTH. (**B**) Calcium. (**C**) Ionized Calcium. (**D**) Phosphate).

4. Discussion

In 1925, in what is known as the first attempted surgical treatment of HPT, Felix Man reported that clinical symptoms improved after removing the enlarged parathyroid gland in a patient with severe bone lesions [26]. With the development of automatic analyzers that can measure serum Ca levels, the detection rate of hypercalcemia has increased since the 1970s. Previously, PHPT was mainly diagnosed by characteristic symptoms, but recently, asymptomatic HPT has been mainly diagnosed through health examinations using biochemical analyses and neck imaging techniques, including ultrasound and computed tomography [27]. According to a single-center experience in Korea, there have been more HPT patients during the last 6 years than during the previous 20 years [28]. This means that the diagnosis rate of HPT patients is increasing in Korea with the development of health screening. However, according to national health insurance data, the incidence of PHPT in Korea is still very low, with an annual incidence ranging from 0.007% to 0.0014% [29]. The most important laboratory tests for HPT are serum Ca and PTH levels. Traditionally, 24 h urine Ca amount, serum chloride concentration, serum chloride-to-P ratio, serum alkaline phosphatase, and tubular P reabsorption rate have been used for differential diagnosis. However, anatomical and functional imaging techniques are more useful in recent clinical diagnosis and surgical planning [30]. Neck ultrasound and 99mTC-sestamibi scanning are very helpful to localize the hyperfunctioning glands, as the sensitivity of ultrasound is 59–89% and that of 99mTC-sestamibi scanning is 54–88% [31–35]. Moreover, the success rate of parathyroid localization increases by 10–20% when both methods are implemented together [36]. The IOPTH assay, which was first proposed in 1988 by Nussbaum et al., is very helpful to confirm the complete removal of the hyperfunctioning parathyroid gland during surgery [37]. It helps the surgeon to evaluate whether bilateral neck exploration is required after focused parathyroidectomy [34,37]. By using localizing imaging techniques and IOPTH, minimally invasive focused parathyroidectomy can be safely achieved for PHPT [36,38,39].

According to the National Institutes of Health guidelines, surgical treatment is recommended for all symptomatic PHPT patients and for asymptomatic patients under 50 years of age who meet the following criteria: 24 h urinary Ca excretion $\leq$ 300 mg, serum Ca more than 1 mg/dL than normal level, creatinine clearance < 30%, or osteoporosis diagnosed by the Dual Energy X-ray Absorptiometry (DEXA) test [40]. The bilateral exploration of the parathyroid glands performed in the past has been recently replaced with focused parathyroidectomy as the standard treatment using localization and IOPTH [10,11,41]. Focused parathyroidectomy has the advantages of a smaller incision, a faster recovery period, and a shorter operation time. In this study, out of the 37 patients in the PHPT group, 35 received focused one-gland parathyroidectomy, 1 received two-gland parathyroidectomy, and 1 received bilateral exploration with the excision of three glands. For the last 2 patients, an extended skin excision was performed because the initial postoperative PTH level did not decrease after the one-gland excision. The PTH level decreased after the additional excision of the parathyroid glands, and the surgery was completed. Thus, our experience suggests that IOPTH is useful in preventing reoperation in some PHPT patients (2/37 (5.4%)).

On the other hand, SHPT is a disease in which the observed parathyroid hyperplasia is due to a damaged kidney. Approximately 5–25% of chronic renal failure patients require surgical treatment of their SHPT [42,43]. According to the Japanese Society of Dialysis Therapy (JSDT) guidelines, the target laboratory ranges for P, Ca, and PTH in SHPT patients are 3.5–6.0 mg/dL, 8.4–10.0 mg/dL, and 60–180 pg/mL, respectively [44]. Pharmacological therapy—vitamin D2, D3 analogs, and calcium preparations—is generally recommended as the initial therapy [41,45–47]. However, the Kidney Disease Outcomes Quality Initiative (KDOQI) guideline recommends that parathyroidectomy should be considered when the PTH level reaches 800 pg/mL, or when the patient suffers from medically intractable hypercalcemia and hyperphosphatemia [48].

Pathologic findings in RHPT are parathyroid hyperplasia in all the parathyroid glands. Thus, the proper removal of the parathyroid glands is very important in RHPT surgery. Therefore, IOPTH is also helpful to confirm the successful removal of the pathologic parathyroid glands in RHPT surgery. Subtotal parathyroidectomy and total parathyroidectomy with autotransplantation are widely used in RHPT surgery. Our center performs subtotal parathyroidectomy due to its several benefits such as shorter operation time, shorter hospital stay, and the reduced need for vitamin D medication [13,14]. As some ectopic parathyroid glands may exist, preoperative images and IOPTH levels are helpful to identify them. In this study, there were no suspected ectopic parathyroid glands in any of the RHPT patients. At the 1-year follow-up, the PTH and Ca levels were stabilized in our patients after the surgery, as shown in Table 3.

Previous studies have reported that the recurrence rate after parathyroidectomy was 0.4% in PHPT and 26% in SHPT [45,49]. In our study, RHPT patients had significantly higher PTH levels before surgery than PHPT patients; the PTH levels were also significantly higher in the IOPTH assay. The PTH levels in the RHPT group were also high at 6 months and 12 months after surgery. However, the postoperative PTH levels met the JSDT guideline in both groups [44]. The preoperative Ca levels were significantly higher in the PHPT group, but postoperative transient hypocalcemia was observed only in the RHPT group 6 months after surgery (9.5 $\pm$ 1.3 mg/dL versus 6.8 $\pm$ 1.2 mg/dL; $p < 0.003$). The Ca levels were recovered to normal levels at 12 months after surgery. This result shows that proper postoperative Ca supplements are required for RHPT patients to prevent symptoms of hypocalcemia due to the excessive bone reabsorption of Ca after parathyroidectomy [13,15].

This study reported on parathyroid surgery for the past 2 years at a single center in South Korea. Considering the low incidence of HPT, our sample of 47 patients was meaningful to evaluate the outcome of parathyroid surgery using IOPTH. Although there are studies suggesting that preoperative localization imaging is a powerful tool for the successful resection of pathologic parathyroid glands in experienced centers, the addition of the IOPTH assay should be considered to prevent surgical failure for novice endocrine surgeons or inexperienced hospitals [50].

Our study was a retrospective descriptive analysis with a small number of patients from a single hospital. Considering the low incidence of HPT, we postulate that this study has clinical significance to the endocrine surgery field, especially in South Korea. Nationwide studies should be conducted in future research to evaluate the outcome of parathyroid surgery in South Korea.

5. Conclusions

HPT is a disease with a low incidence rate, but the frequency of detection is increasing with the development of medical examination techniques. Focused parathyroidectomy with PHPT is considered acceptable. In RHPT, it is necessary to completely remove the hyperfunctioning gland. The IOPTH assay can help surgeons to reduce incomplete excision and recurrence.

Author Contributions: Conceptualization, S.H.K., Y.S.C. and J.W.Y.; Data curation, S.Y.L., E.A.M. and J.W.Y.; Formal analysis, S.Y.L. and J.W.Y.; Funding acquisition, J.W.Y.; Investigation, S.Y.L. and J.W.Y.; Methodology, Y.M.H., Y.S.C. and J.W.Y.; Project administration, Y.S.C. and S.Y.L.; Resource acquisition, Y.S.C. and J.W.Y.; Supervision, J.W.Y.; Writing—original draft, Y.S.C.; Writing—review and editing, S.H.K. and J.W.Y. All authors have read and agreed to the published version of the manuscript.

Funding: This research received no external funding.

Institutional Review Board Statement: This study was approved by the Institutional Review Board of Inha University Hospital on 30 May 2022 (IRB number: INH 2022-05-028).

Informed Consent Statement: Patient consent was waived as this study is retrospective and had no additional interventions and collected no collecting private information.

Data Availability Statement: The data presented in this study are available in this article.

Acknowledgments: This work was supported by an Inha University Hospital research grant.

Conflicts of Interest: The authors have no conflict of interest to declare.

References

1. Fraser, W.D. Hyperparathyroidism. *Lancet* **2009**, *374*, 145–158. [CrossRef]
2. Cho, N.L.; Doherty, G.M. 55—Principles in surgical management of primary hyperparathyroidism. In *Surgery of the Thyroid and Parathyroid Glands*, 3rd ed.; Randolph, G.W., Ed.; Elsevier: Amsterdam, The Netherlands, 2021; pp. 502–516.e5.
3. Tominaga, Y. Surgical management of secondary and tertiary hyperparathyroidism. In *Surgery of the Thyroid and Parathyroid Glands*, 3rd ed.; Randolph, G.W., Ed.; Elsevier: Amsterdam, The Netherlands, 2021; pp. 564–575.e4.
4. Cunningham, J.; Locatelli, F.; Rodriguez, M. Secondary hyperparathyroidism: Pathogenesis, disease progression, and therapeutic options. *Clin. J. Am. Soc. Nephrol.* **2011**, *6*, 913–921. [CrossRef] [PubMed]
5. Madorin, C.; Owen, R.P.; Fraser, W.D.; Pellitteri, P.K.; Radbill, B.; Rinaldo, A.; Seethala, R.R.; Shaha, A.R.; Silver, C.E.; Suh, M.Y.; et al. The surgical management of renal hyperparathyroidism. *Eur. Arch. Otorhinolaryngol.* **2012**, *269*, 1565–1576. [CrossRef] [PubMed]
6. Adler, J.T.; Sippel, R.S.; Chen, H. New trends in parathyroid surgery. *Curr. Probl. Surg.* **2010**, *47*, 958–1017. [CrossRef]
7. Wermers, R.A.; Khosla, S.; Atkinson, E.J.; Achenbach, S.J.; Oberg, A.L.; Grant, C.S.; Melton, L.J., 3rd. Incidence of primary hyperparathyroidism in Rochester, Minnesota, 1993–2001: An update on the changing epidemiology of the disease. *J. Bone Miner. Res.* **2006**, *21*, 171–177. [CrossRef]
8. Macfarlane, D.P.; Yu, N.; Donnan, P.T.; Leese, G.P. Should 'mild primary hyperparathyroidism' be reclassified as 'insidious': Is it time to reconsider? *Clin. Endocrinol.* **2011**, *75*, 730–737. [CrossRef]

9.   Marcocci, C.; Cetani, F. Primary hyperparathyroidism. *N. Engl. J. Med.* **2011**, *365*, 2389–2397. [CrossRef]
10. Dillavou, E.D.; Cohn, H.E. Minimally invasive parathyroidectomy: 101 consecutive cases from a single surgeon. *J. Am. Coll. Surg.* **2003**, *197*, 1–7. [CrossRef]
11. Grant, C.S.; Thompson, G.; Farley, D.; van Heerden, J. Primary hyperparathyroidism surgical management since the introduction of minimally invasive parathyroidectomy: Mayo Clinic experience. *Arch. Surg.* **2005**, *140*, 472–479. [CrossRef]
12. Park, J.H.; Kang, S.W.; Jeong, J.J.; Nam, K.H.; Chang, H.S.; Chung, W.Y.; Park, C.S. Surgical treatment of tertiary hyperparathyroidism after renal transplantation: A 31-year experience in a single institution. *Endocr. J.* **2011**, *58*, 827–833. [CrossRef]
13. Choi, H.R.; Aboueisha, M.A.; Attia, A.S.; Omar, M.; ELnahla, A.; Toraih, E.A.; Shama, M.; Chung, W.Y.; Jeong, J.J.; Kandil, E. Outcomes of subtotal parathyroidectomy versus total parathyroidectomy with autotransplantation for tertiary hyperparathyroidism: Multi-institutional study. *Ann. Surg.* **2021**, *274*, 674–679. [CrossRef] [PubMed]
14. Yuan, Q.; Liao, Y.; Zhou, R.; Liu, J.; Tang, J.; Wu, G. Subtotal parathyroidectomy versus total parathyroidectomy with autotransplantation for secondary hyperparathyroidism: An updated systematic review and meta-analysis. *Langenbecks Arch. Surg.* **2019**, *404*, 669–679. [CrossRef] [PubMed]
15. Choi, Y.-S.; Lee, K.E.; Koo, D.H.; Oh, E.M.; Choi, J.Y.; Park, K.-W.; Noh, D.Y.; Youn, Y.-K.; Oh, S.-K. The postoperative outcomes of patients with primary, secondary and tertiary hyperparathyroidism: 14 year experience of Seoul National University Hospital. *Korean J. Clin. Oncol.* **2011**, *7*, 52–59. [CrossRef]
16. Jeon, H.J.; Kim, Y.J.; Kwon, H.Y.; Koo, T.Y.; Baek, S.H.; Kim, H.J.; Huh, W.S.; Huh, K.H.; Kim, M.S.; Kim, Y.S.; et al. Impact of parathyroidectomy on allograft outcomes in kidney transplantation. *Transpl. Int.* **2012**, *25*, 1248–1256. [CrossRef] [PubMed]
17. Kim, M.S.; Kim, G.H.; Lee, C.H.; Park, J.S.; Lee, J.Y.; Tae, K. Surgical outcomes of subtotal parathyroidectomy for renal hyperparathyroidism. *Clin. Exp. Otorhinolaryngol.* **2020**, *13*, 173–178. [CrossRef]
18. Kim, W.W.; Rhee, Y.; Ban, E.J.; Lee, C.R.; Kang, S.W.; Jeong, J.J.; Nam, K.H.; Chung, W.Y.; Park, C.S. Is focused parathyroidectomy appropriate for patients with primary hyperparathyroidism? *Ann. Surg. Treat. Res.* **2016**, *91*, 97–103. [CrossRef] [PubMed]
19. Kim, W.Y.; Lee, J.B.; Kim, H.Y.; Woo, S.U.; Son, G.S.; Bae, J.W. Achievement of the National Kidney Foundation Kidney Disease Outcomes Quality Initiative: Recommended serum calcium, phosphate and parathyroid hormone values with parathyroidectomy in patients with secondary hyperparathyroidism. *J. Korean Surg. Soc.* **2013**, *85*, 25–29. [CrossRef]
20. Kim, Y.I.; Jung, Y.H.; Hwang, K.T.; Lee, H.Y. Efficacy of ^{99m}Tc-sestamibi SPECT/CT for minimally invasive parathyroidectomy: Comparative study with ^{99m}Tc-sestamibi scintigraphy, SPECT, US and CT. *Ann. Nucl. Med.* **2012**, *26*, 804–810. [CrossRef]
21. Cho, E.; Chang, J.M.; Yoon, S.Y.; Lee, G.T.; Ku, Y.H.; Kim, H.I.; Lee, M.C.; Lee, G.H.; Kim, M.J. Preoperative localization and intraoperative parathyroid hormone assay in Korean patients with primary hyperparathyroidism. *Endocrinol. Metab.* **2014**, *29*, 464–469. [CrossRef]
22. Kim, H.G.; Kim, W.Y.; Woo, S.U.; Lee, J.B.; Lee, Y.M. Minimally invasive parathyroidectomy with or without intraoperative parathyroid hormone for primary hyperparathyroidism. *Ann. Surg. Treat. Res.* **2015**, *89*, 111–116. [CrossRef]
23. Kim, W.Y.; Lee, J.B.; Kim, H.Y. Efficacy of intraoperative parathyroid hormone monitoring to predict success of parathyroidectomy for secondary hyperparathyroidism. *J. Korean Surg Soc.* **2012**, *83*, 1–6. [CrossRef] [PubMed]
24. Paek, S.H.; Kim, S.J.; Choi, J.Y.; Lee, K.E. Clinical usefulness of intraoperative parathyroid hormone monitoring for primary hyperparathyroidism. *Ann. Surg. Treat. Res.* **2018**, *94*, 69–73. [CrossRef] [PubMed]
25. Carneiro, D.M.; Solorzano, C.C.; Nader, M.C.; Ramirez, M.; Irvin, G.L., 3rd. Comparison of intraoperative iPTH assay (QPTH) criteria in guiding parathyroidectomy: Which criterion is the most accurate? *Surgery* **2003**, *134*, 973–979; discussion 979–981. [CrossRef] [PubMed]
26. Liu, H.; Wilkerson, M.L.; Lin, F. Thyroid, parathyroid, and adrenal glands. In *Handbook of Practical Immunohistochemistry: Frequently Asked Questions*; Lin, F., Prichard, J.W., Liu, H., Wilkerson, M.L., Eds.; Springer International Publishing: Cham, Switzerland, 2022; pp. 339–374.
27. Bilezikian, J.P.; Khan, A.A.; Potts, J.T., Jr. Guidelines for the management of asymptomatic primary hyperparathyroidism: Summary statement from the third international workshop. *J. Clin. Endocrinol. Metab.* **2009**, *94*, 335–339. [CrossRef]
28. Kwon, W.; Jang, M.C.; Noh, D.Y.; Youn, Y.K.; Oh, S.K. Primary hyperparathyroidism: A 26-year experience at Seoul National University Hospital. *Korean J. Endocr. Surg.* **2007**, *7*, 147–154. [CrossRef]
29. Kim, J.K.; Chai, Y.J.; Chung, J.K.; Hwang, K.T.; Heo, S.C.; Kim, S.J.; Choi, J.Y.; Yi, K.H.; Kim, S.W.; Cho, S.Y.; et al. The prevalence of primary hyperparathyroidism in Korea: A population-based analysis from patient medical records. *Ann. Surg. Treat. Res.* **2018**, *94*, 235–239. [CrossRef]
30. Heath, D.A. Primary hyperparathyroidism. Clinical presentation and factors influencing clinical management. *Endocrinol. Metab. Clin. N. Am.* **1989**, *18*, 631–646. [CrossRef]
31. Arici, C.; Cheah, W.K.; Ituarte, P.H.; Morita, E.; Lynch, T.C.; Siperstein, A.E.; Duh, Q.Y.; Clark, O.H. Can localization studies be used to direct focused parathyroid operations? *Surgery* **2001**, *129*, 720–729. [CrossRef]
32. Ryan, J.A., Jr.; Eisenberg, B.; Pado, K.M.; Lee, F. Efficacy of selective unilateral exploration in hyperparathyroidism based on localization tests. *Arch. Surg.* **1997**, *132*, 886–890; discussion 890–891. [CrossRef]
33. Martin, R.C., 2nd; Greenwell, D.; Flynn, M.B. Initial neck exploration for untreated hyperparathyroidism. *Am. Surg.* **2000**, *66*, 269–272.
34. Irvin, G.L., 3rd; Dembrow, V.D.; Prudhomme, D.L. Operative monitoring of parathyroid gland hyperfunction. *Am. J. Surg.* **1991**, *162*, 299–302. [CrossRef]

35. Chapuis, Y.; Fulla, Y.; Bonnichon, P.; Tarla, E.; Abboud, B.; Pitre, J.; Richard, B. Values of ultrasonography, sestamibi scintigraphy, and intraoperative measurement of 1-84 PTH for unilateral neck exploration of primary hyperparathyroidism. *World J. Surg.* **1996**, *20*, 835–839; discussion 839–840. [CrossRef] [PubMed]

36. Sugg, S.L.; Krzywda, E.A.; Demeure, M.J.; Wilson, S.D. Detection of multiple gland primary hyperparathyroidism in the era of minimally invasive parathyroidectomy. *Surgery* **2004**, *136*, 1303–1309. [CrossRef]

37. Nussbaum, S.R.; Thompson, A.R.; Hutcheson, K.A.; Gaz, R.D.; Wang, C.A. Intraoperative measurement of parathyroid hormone in the surgical management of hyperparathyroidism. *Surgery* **1988**, *104*, 1121–1127. [PubMed]

38. Haciyanli, M.; Lal, G.; Morita, E.; Duh, Q.Y.; Kebebew, E.; Clark, O.H. Accuracy of preoperative localization studies and intraoperative parathyroid hormone assay in patients with primary hyperparathyroidism and double adenoma. *J. Am. Coll. Surg.* **2003**, *197*, 739–746. [CrossRef]

39. Fraker, D.L.; Harsono, H.; Lewis, R. Minimally invasive parathyroidectomy: Benefits and requirements of localization, diagnosis, and intraoperative PTH monitoring. Long-term results. *World J. Surg.* **2009**, *33*, 2256–2265. [CrossRef]

40. Eigelberger, M.S.; Cheah, W.K.; Ituarte, P.H.; Streja, L.; Duh, Q.Y.; Clark, O.H. The NIH criteria for parathyroidectomy in asymptomatic primary hyperparathyroidism: Are they too limited? *Ann. Surg.* **2004**, *239*, 528–535. [CrossRef]

41. Aygün, N.; Uludağ, M. Surgical treatment of primary hyperparathyroidism: Which therapy to whom? *Sisli Etfal Hastan. Tip Bul.* **2019**, *53*, 201–214. [CrossRef]

42. Glassford, D.M.; Remmers, A.R., Jr.; Sarles, H.E.; Lindley, J.D.; Scurry, M.T.; Fish, J.C. Hyperparathyroidism in the maintenance dialysis patient. *Surg. Gynecol. Obstet.* **1976**, *142*, 328–332.

43. Higgins, R.M.; Richardson, A.J.; Ratcliffe, P.J.; Woods, C.G.; Oliver, D.O.; Morris, P.J. Total parathyroidectomy alone or with autograft for renal hyperparathyroidism? *Q. J. Med.* **1991**, *79*, 323–332.

44. Kazama, J.J. Japanese Society of Dialysis Therapy treatment guidelines for secondary hyperparathyroidism. *Ther. Apher. Dial.* **2007**, *11* (Suppl. 1), S44–S47. [CrossRef] [PubMed]

45. Stracke, S.; Keller, F.; Steinbach, G.; Henne-Bruns, D.; Wuerl, P. Long-term outcome after total parathyroidectomy for the management of secondary hyperparathyroidism. *Nephron Clin. Pract.* **2009**, *111*, c102–c109. [CrossRef] [PubMed]

46. Tominaga, Y.; Matsuoka, S.; Uno, N.; Sato, T. Parathyroidectomy for secondary hyperparathyroidism in the era of calcimimetics. *Ther. Apher. Dial.* **2008**, *12* (Suppl. 1), S21–S26. [CrossRef] [PubMed]

47. Saliba, W.; El-Haddad, B. Secondary hyperparathyroidism: Pathophysiology and treatment. *J. Am. Board Fam. Med.* **2009**, *22*, 574–581. [CrossRef]

48. National Kidney Foundation. K/DOQI clinical practice guidelines for bone metabolism and disease in chronic kidney disease. *Am. J. Kidney Dis.* **2003**, *42*, S1–S201. [CrossRef]

49. Clark, O.H.; Wilkes, W.; Siperstein, A.E.; Duh, Q.Y. Diagnosis and management of asymptomatic hyperparathyroidism: Safety, efficacy, and deficiencies in our knowledge. *J. Bone Miner. Res.* **1991**, *6* (Suppl. 2), S135–S142; discussion 151–152. [CrossRef]

50. Patel, N.; Mihai, R. Long-term cure of primary hyperparathyroidism after scan-directed parathyroidectomy: Outcomes from a UK endocrine surgery unit. *World J. Surg.* **2022**, *46*, 2189–2194. [CrossRef]

Article

The Utility of 4D-CT Imaging in Primary Hyperparathyroidism Management in a Low-Volume Center

Marko Murruste [1,*], Martin Kivilo [2], Karri Kase [1,2], Ülle Kirsimägi [1], Annika Tähepõld [3] and Kaia Tammiksaar [4]

1 Surgery Clinic of Tartu University Hospital, 50406 Tartu, Estonia; karri.kase@kliinikum.ee (K.K.); ylle.kirsimagi@kliinikum.ee (Ü.K.)
2 Faculty of Medicine, University of Tartu, 50406 Tartu, Estonia; martinkivilo@gmail.com
3 Radiology Clinic of Tartu University Hospital, 50406 Tartu, Estonia; annika.tahepold@kliinikum.ee
4 Internal Medicine Clinic of Tartu University Hospital, 50406 Tartu, Estonia; kaia.tammiksaar@kliinikum.ee
* Correspondence: marko.murruste@kliinikum.ee

Abstract: *Background:* Ultrasonography (US) and the 99mTc-sestamibi parathyroid scan (SPS) may have suboptimal accuracy when detecting the localization of enlarged parathyroid gland(s) (PTG). Therefore, the more accurate four-dimensional computed tomography scan (4D-CT) has been employed for PTG imaging. Currently, there is a paucity of data evaluating the utility of 4D-CT in low caseload settings. *Aim and Objectives:* To evaluate the impact of PTG imaging, using 4D-CT in conjunction with its intraoperatively displayed results, on the outcomes of surgical PTX. *Materials and Methods:* A single-center retrospective analysis of surgically treated patients with pHPT from 01/2010 to 01/2021 was conducted. An evaluation of the impact of the preoperative imaging modalities on the results of surgical treatment was carried out. *Results:* During the study period, 290 PTX were performed; 45 cases were excluded due to surgery for secondary, tertiary or recurrent HPT, or due to the use of alternative imaging techniques. The remaining 245 patients were included in the study. US was carried out for PTG imaging in 236 (96.3%), SPS in 93 (38.0%), and 4D-CT in 52 patients (21.2%). The use of 4D-CT was associated with a significantly higher rate of successful localization of enlarged PTG (49 cases, 94.2%) compared to US and SPS (74 cases, 31.4%, and 54 cases, 58.1%, respectively). We distinguished between three groups of patients based on preoperative imaging: (1) PTG lateralization via US or SPS in 106 (43.3%) cases; (2) precise localization of PTG via 4D-CT in 49 (20.0%) patients; and (3) in 90 cases (36.7%), PTG imaging failed to localize an enlarged gland. The group of 4D-CT localization had significantly shorter operative time, lower rate of simultaneous thyroid resections, as well as lower rate of removal of ≥2 PTG, compared to the other groups. The 4D-CT imaging was also associated with the lowest perioperative morbidity and with the lowest median PTH in the one month follow-up; however, compared to the other groups, these differences were statistically not significant. The implementation of 4D-CT (since 01/2018) was associated with a decrease in the need for redo surgery (from 11.5% to 7.3%) and significantly increased the annual case load of PTX at our institution (from 15.3 to 41.0) compared to the period before 4D-CT diagnostics. *Conclusions:* 4D-CT imaging enabled to precisely locate almost 95% of enlarged PTG in patients with pHPT. Accurate localization and intraoperatively displayed imaging results are useful guides for surgeons to make PTX a faster and safer procedure in a low-volume center.

Keywords: primary hyperparathyroidism; parathyroid imaging; parathyroidectomy; four-dimensional computed tomography

Citation: Murruste, M.; Kivilo, M.; Kase, K.; Kirsimägi, Ü.; Tähepõld, A.; Tammiksaar, K. The Utility of 4D-CT Imaging in Primary Hyperparathyroidism Management in a Low-Volume Center. *Medicina* **2023**, *59*, 1415. https://doi.org/10.3390/medicina59081415

Academic Editor: Jin Wook Yi

Received: 1 July 2023
Revised: 24 July 2023
Accepted: 1 August 2023
Published: 3 August 2023

1. Introduction

Primary hyperparathyroidism (pHPT) is an increasingly common endocrine disorder, reaching a prevalence of 0.3% in the general European population [1,2]. Furthermore, in the Scandinavian countries, the prevalence of pHPT has reached 2–5% in peri- and postmenopausal women [3]. The wider availability of blood calcium screening tests and ultrasonography (US) has probably led to this increase in the recent decades [4]. As surgical treatment is the only definitive cure for pHPT; a parallel increase in the prevalence of parathyroidectomies (PTX) has been seen in the past decades [5,6].

When compared to conservative management, PTX is a relatively safe and cost-effective procedure [7]. Cure rates of surgical PTX exceed 95% at centers with high expertise [5]. Although this procedure is associated with a relatively low risk of complications, the ultimate goal of surgery—normal function of the remaining parathyroid glands (PTG) manifesting as persistent normocalcemia—is not always met. This is often due to difficulties in the preoperative and intraoperative localization of pathologic PTG, leading to the persistence of pHPT and the need for redo surgery [8]. Although the need for preoperative PTG imaging has been a subject of numerous debates, it has been demonstrated that the preoperative localization of enlarged PTG facilitates the intraoperative exploration of the culprit gland. Furthermore, PTG imaging also has the potential to prevent a large number of bilateral neck explorations (BNE) and promote unilateral neck exploration or minimally invasive PTX instead [9]. Therefore, the current guidelines recommend the preoperative localization of hyperfunctioning PTG for the selection of suspect PTX candidates, as it allows for a more focused approach, reduces operation time and complications, and results in a better cosmetic outcome and greater patient satisfaction [10,11]. Although the preferred sequence of imaging continues to evolve and there exists no universally agreed algorithm, US is usually the first line modality, followed in some institutions by ^{99m}Tc sestamibi parathyroid scan with subtraction imaging (SPS). However, both tests may have suboptimal localizing accuracy with a considerable rate of failed localizations. Therefore, a more accurate four-dimensional computed tomography scan (4D-CT) has been suggested for PTG imaging [12].

Another aspect affecting successful PTX and biochemical cure is surgeons' experience. Multiple studies have reported better outcomes of parathyroid surgery in high-volume centers, however, no definite volume threshold has been established [13]. Although some studies have proposed a threshold of 20 PTX per year, this number is purely based on professional opinions, a so-called 'pragmatic and achievable target' [14,15].

The above-described problems were also seen in our center. We had a relatively high percentage of patients with failed preoperative localization of PTG, and our annual caseload was relatively low (ca 10–15 cases of surgically treated patients with pHPT) [16]. Thus, in 2018, we attempted to improve the PTX success rate by implementing 4D-CT imaging for the localization of pathologically enlarged PTG from 2018.

The aim of the present study was to evaluate the impact of using 4D-CT as the PTG imaging modality on the outcomes of surgical PTX.

Our primary outcome was sensitivity of 4D-CT in comparison to US and SPS for preoperative PTG adenoma localization. The secondary outcomes were the need for redo surgery and morbidity. Additionally, the characteristics of surgical PTX, i.e., operative time, number of removed parathyroid glands, and rate of simultaneous thyroid surgery, were assessed.

We hypothesized that the success rate of 4D-CT in detecting enlarged PTG would be higher compared to US and SPS, which would result in better outcomes of surgical treatment. Furthermore, we hoped that the implementation of 4D-CT would make PTX outcomes comparable to those of large volume centers.

2. Methods

Ethics. The present study is a single-center retrospective analysis of surgically treated patients with pHPT. The Research Ethics Committee of the University of Tartu approved this clinical research (protocol 300/T-4).

Patients. All adult patients aged 18 years or older, who were operated due to pHPT at the Department of Surgery of Tartu University Hospital between 01/2010 and 01/2021, and on whom US, SPS, or 4D-CT for PTG imaging was used, were included in the study. The exclusion criteria were surgery for secondary, tertiary, or recurrent HPT. We also excluded a few patients whose PTG imaging investigations were magnetic resonance imaging (MRI) or conventional computed tomography (CT).

During the study period, 290 parathyroid operations were carried out; 32 cases were excluded due to surgery for secondary, tertiary, or recurrent HPT and 13 cases due to PTG imaging via MRI or conventional CT. The remaining 245 patients, who were operated for pHPT and whose imaging investigations were US, SPS, or 4D-CT, were included in the study (Figure 1).

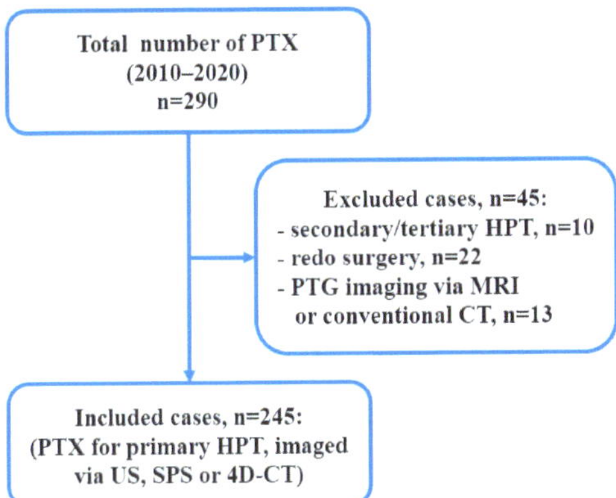

Figure 1. Study flowchart. PTX parathyroidectomy; HPT hyperparathyroidism; PTG parathyroid gland; US ultrasonography; SPS sestamibi parathyroid scan; 4D-CT four-dimensional computed tomography; MRI magnetic resonance imaging.

PTG imaging. In the present study, the routine management of patients with suspected pHPT consisted of preoperative biochemical confirmation of pHPT, followed by localization investigations, including US (performed by endocrinologists) and SPS in cases of inconclusive US imaging. Since 2018, we have replaced SPS with 4D-CT imaging, and SPS was only used in cases where 4D-CT was contraindicated (e.g., renal insufficiency, intolerance of iodine based contrast media). The 4D-CT scanning protocol consists of precontrast imaging followed by arterial (25 to 30 s after contrast injection), early-delayed (40 to 45 s after contrast injection), and late-delayed (80 to 90 s after contrast injection) phases.

To facilitate better intraoperative localization of enlarged PTG, the results of 4D-CT imaging (precise localization of suspected pathological PTG) were always displayed intraoperatively using the axial, coronal, and/or sagittal CT-planes.

Surgical treatment consisted of a small Kocher incision (usually up to 5 cm) for access, followed by the classical medial approach to explore the thyroid and parathyroid regions, mobilizing the strap muscles from their midline position by progressive lateral mobilization.

For patients with failed preoperative localization or with multiglandular PTG involvement, a classical BNE was carried out. In the case of successful lateralization of enlarged PTG using US or SPS, a unilateral neck exploration was employed. For patients with precise preoperative localization of pathologic PTG via 4D-CT imaging, a one-region dissection in the suspected area was undertaken. In the case of failed focused exploration of PTG, conversion to classical BNE was employed.

Database. The study database comprises data about surgical treatment (operative time, number of removed PTG, simultaneous thyroid surgery, complications, and length of stay) and biochemical characteristics before and after surgical treatment (serum calcium, ionized calcium, parathyroid hormone (PTH), 25-hydroxyvitamin D, phosphate, and creatinine levels). Biochemical follow-up data were obtained one month and one year after surgical treatment.

Persistence of pHPT was defined as the presence of one of the following: reoperation for pHPT during 2 years of follow-up, or ionized calcium levels above 1.32 mmol/L with synchronous PTH levels over 6.9 pmol/L.

Data analysis. The impact of the preoperative imaging modalities on the results of surgical treatment were evaluated in three groups: (1) PTG localization via US or SPS, as these modalities enabled to detect the affected side ('lateralization'), however, precise three-dimensional localization was not employed; (2) precise localization of PTG via 4D-CT; and (3) cases where preoperative PTG localization was not successful ('image negative patients').

An additional assessment of the results was carried out in two time frames: before the implementation of 4D-CT (from 01/2010 to 12/2017) and after it (01/2018 to 01/2021).

Statistical analysis. All perioperative and follow-up data were entered into a computerized database (Microsoft Access 2016). The software package Statistica version 13.3 (TIBCO Software, Palo Alto, CA, USA) was used for statistical calculations.

The main characteristics are presented as medians (with 25% and 75% percentiles) in the case of non-normal distribution of variables; dichotomous variables are reported as counts and percentages. Comparisons between the study groups were made using the following tests: Fisher's exact test in the case of percentages, non-parametric Mann–Whitney U-test in the case of medians for non-normally distributed variables. The level of statistical significance was set at $p < 0.05$.

3. Results

Patients. During the study period, 245 patients who were operated for pHPT and whose imaging investigations were US, SPS, or 4D-CT were included in the study.

Preoperative data. The median age of the patients at the time of surgical treatment was 66 years; there were 218 female (89.0%) and 27 male patients (Table 1). The preoperative biochemical characteristics were: median ionized calcium level 1.51 mmol/L; PTH 15.0 pmol/L; and phosphate 0.86 mmol/L.

Table 1. Preoperative characteristics of the patients.

Characteristic	Patients, n = 245
Age (y)	66 (58–74)
Female gender, n (%)	218 (89.0%)
iCa *	1.51 (1.44–1.61)
PTH [#]	15.0 (10.4–25.3)
Phosphate [□]	0.86 (0.73–0.95)
Parathyroid imaging:	
— US, n (%)	236 (96.3)
— successful localization, n (%)	74 (31.4)
— SPS, n (%)	93 (38.0)
— successful localization, n (%)	54 (58.1)
— 4D-CT, n (%)	52 (21.2)
— successful localization, n (%)	49 (94.2)

Data are presented as median (with interquartile range) or percentages. Y years; iCa ionized calcium; PTH parathyroid hormone; US ultrasonography; SPS sestamibi parathyroid scan; 4D-CT four-dimensional computed tomography. * Reference value 1.16–1.32 mmol/L, [#] Reference value 1.6–6.9 pmol/L, [□] Reference value 0.81–1.45 mmol/L.

PTG imaging. For PTG imaging, US was carried out in 236 (96.3%, Figure 2), SPS in 93 (38.0%), and 4D-CT in 52 patients (21.2%) since its implementation in 2018. There occurred no complications associated with the use of 4D-CT imaging (e.g., development or exacerbation of renal insufficiency; intolerance of intravenous iodine based contrast media; etc.). However, for 6 patients, 4D-CT was contraindicated due to an underlying renal insufficiency or known allergy to iodine-based contrast media.

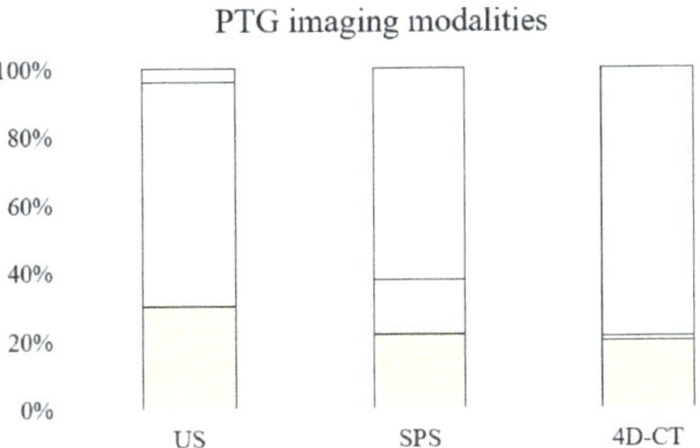

Figure 2. Use of PTG imaging modalities in 245 patients. Image-negative cases (striped yellow area). Successful localization of enlarged PTG (yellow area). *PTG* parathyroid gland; *US* ultrasonography; *SPS* sestamibi parathyroid scan; *4D-CT* four-dimensional computed tomography.

The use of 4D-CT was associated with a significantly higher rate of successful localization of enlarged PTG (49 cases, 94.2%) compared to US and SPS (74 cases, 31.4%; and 54 cases, 58.1%, respectively).

Results of surgical treatment.

According to the preoperative data gathered on PTG localization diagnostics, three groups of patients were distinguished: (1) PTG lateralization (via UH or SPS), (2) precise PTG localization (via 4D-CT), and (3) patients with non-successful localization of PTG. There were 106 patients (43.3%) in the first group, 49 in the second (20.0%), and 90 (36.7%) in the third (Table 2).

More precise localizations of PTG with the use of 4D-CT led to significantly shorter median operative time, this being 35 min (Figure 3) compared to 51 min if PTG localization was not successful and 53 min after localization via US/SPS.

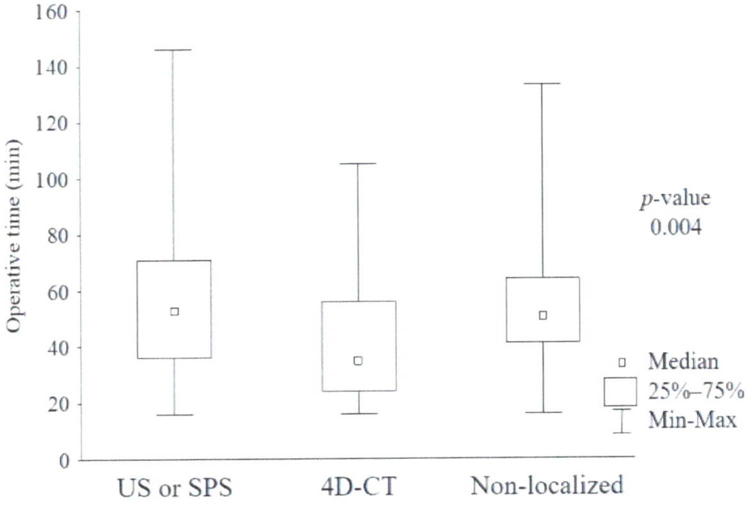

Figure 3. Operative time of parathyroidectomy according to preoperative localization diagnostics. *US* ultrasonography; *SPS* sestamibi parathyroid scan; *4D-CT* four-dimensional computed tomography.

An assessment of the characteristics of PTX revealed a significantly higher proportion of patients in the group of non-localized PTG undergoing removal of ≥ 2 PTG (35.6%) compared to 23.6% and 14.3% in the US/SPS and 4D-CT groups, respectively. Although the rate of true histologically proven multiglandular disease was similar between the groups, it is obvious that in the remaining cases, 'innocent' PTG were removed.

The unsuccessful preoperative localization of PTG was also associated with a significantly higher rate of simultaneous thyroid resections (52.2%) compared to the US/SPS (28.3%) and 4D-CT (28.6%) groups.

Moreover, 4D-CT imaged patients showed the lowest complications rate at 2.0%, compared to 13.2% and 10.0% for the US/SPS imaged patients and the patients with failed preoperative imaging, respectively. However, the difference between the groups remained statistically non-significant ($p = 0.076$).

A short-term biochemical follow-up (one month after surgery) revealed the lowest PTH and ionized calcium levels to be in the group of 4D-CT imaged cases. Nonetheless, compared to the US/SPS imaged and non-localized cases, the difference between the groups was not significant ($p = 0.082$ for PTH, and $p = 0.318$ for ionized calcium).

Persistence of pHPT was diagnosed in three patients (6.1%) of the 4D-CT group, and in 13 (12.3%) and 7 patients (7.8%) of the US/SPS and the non-localized groups, respectively. A reoperation for persistence of hyperparathyroidism was undertaken in most cases during the first year after initial surgery. As a result of successful reoperation, a one-year biochemical follow-up demonstrated similar ionized calcium and PTH levels in all three groups.

Table 2. Comparison of operative and follow-up characteristics of the patients according to the results of preoperative imaging.

	Localized via US or SPS (106)	Localized via 4D-CT (49)	Non-Localized PTG (90)	*p*-Value
Operative characteristics:				
— Operative time (min)	53 (36–71)	35 (24–56)	51 (41–64)	**0.004**
— Removed ≥2 PTG, n (%)	25 (23.6)	7 (14.3)	32 (35.6)	**0.018**
— True multiglandular disease &, n (%)	17 (16.0)	5 (10.2)	16 (17.8)	0.505
— PTG size (mm)	15 (10–20)	13 (10–19)	15 (8–20)	0.728
— Simultaneous thyroid surgery, n (%)	30 (28.3)	14 (28.6)	47 (52.2)	**0.001**
Complications, n (%)	14 (13.2)	1 (2.0)	9 (10.0)	0.076
— Wound hematoma, n	0	0	3	-
— RLN injury, n	3	1	1	-
— Subjective dysphonia, n	5	0	0	-
— Other	6	0	5	-
Length of stay (d)	2 (2–2)	2 (2–2)	2 (2–2)	0.272
Follow-up (30 day):				
— iCa *	1.29 (1.23–1.35)	1.25 (1.22–1.32)	1.29 (1.23–1.33)	0.318
— PTH #	7.1 (5.1–12.3)	5.6 (4.3–7.6)	6.2 (3.8–11.0)	0.082
Follow-up (1 year):				
— iCa *	1.27 (1.23–1.30)	1.25 (1.20–1.32)	1.26 (1.22–1.33)	0.798
— PTH #	5.4 (4.5–9.9)	5.7 (3.8–8.3)	5.4 (3.7–9.5)	0.735
Redo surgery, n (%)	13 (12.3)	3 (6.1)	7 (7.8)	0.467

Data are presented as median (with interquartile range) or percentages. US ultrasonography; SPS sestamibi parathyroid scan; 4D-CT four-dimensional computed tomography; PTG parathyroid gland; RLN recurrent laryngeal nerve; iCa ionized calcium; PTH parathyroid hormone. & Histologically proven multiglandular parathyroid gland involvement, * Reference value 1.16–1.32 mmol/L, # Reference value 1.6–6.9 pmol/L.

Next, we evaluated the prevalence of complications and failure to cure as a combined metric of negative outcome suggested by Ullmann et al. [8]. There was no significant difference between the groups. Yet, the lowest rate of one or both negative events occurred in the group of 4D-CT localized PTG, this being 8.2% compared to 23.6% and 16.7% in the groups of UH/SPS localized and of non-localized patients, respectively.

Additionally, an analysis of the results for two time frames was performed. During the first period (from 2010 to 2017), PTG was localized via US or SPS; the second period (from 2018 to 2020) started after the implementation of 4D-CT diagnostics, which was indicated for patients with inconclusive UH results. This analysis revealed two important associations with the implementation of 4D-CT (Figure 4). The need for redo surgery fell from 11.5% during the first period to 7.3% during the second period, and the annual case-load of PTX almost tripled at our institution from 15.3 cases per year for the first period to 41 cases per year for the second period.

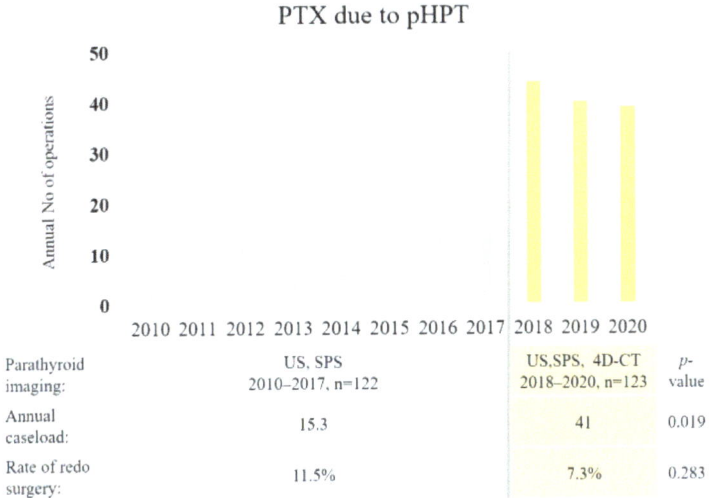

Parathyroid imaging:	US, SPS 2010–2017, n=122	US,SPS, 4D-CT 2018–2020, n=123	*p*-value
Annual caseload:	15.3	41	0.019
Rate of redo surgery:	11.5%	7.3%	0.283

Figure 4. Two periods of imaging: before and after implementation of 4D-CT scan for PTG localization. Impact on the annual PTX case-load and on the rate of redo surgery. *PTX* parathyroidectomy; *pHPT* primary hyperparathyroidism; *US* ultrasonography; *SPS* sestamibi parathyroid scan; *4D-CT* four-dimensional computed tomography.

4. Discussion

The present retrospective study reports an impact of the implementation of 4D-CT imaging of PTG on the results of surgical treatment of pHPT in a low-volume center.

The need for PTG imaging has been a subject of debate over decades. It has been suggested repeatedly that an experienced parathyroid surgeon is capable of localizing a vast majority of parathyroid glands without imaging assistance [17]. It has even been stated that the greatest challenge in the preoperative localization of PTG adenoma is locating an experienced parathyroid surgeon [11]. However, as parathyroid surgery also has to be performed at low or medium case-load institutions, expert surgeons are not always available [18]. Things are further complicated when we consider that this experience can only be gained from parathyroid surgery. It has been demonstrated that the surgical experience gained from non-parathyroid surgery in the same anatomical region (e.g., thyroid surgery) does not improve outcomes in parathyroid surgery. For example, centers performing a large number of thyroid but low number of parathyroid surgical procedures have a significantly higher rate of reoperative parathyroid surgery [19]. Thus, other methods need to compliment the surgeon's experience.

Preoperativne imaging is a good way to compensate for the experience cap in lower volume centers. Frank et al. demonstrated that negative preoperative imaging correlates with reduced success rates of surgical exploration and a higher need for redo surgery [20]. An even more convincing argument for the use of extended localization diagnostics is that focused dissection in the area of enlarged PTG is associated with reduced morbidity [21]. Therefore, as the precise preoperative anatomical localization of enlarged PTG facilitates the intraoperative exploration of the culprit gland, PTG imaging has the potential to prevent a large number of BNE, as well as promote unilateral neck exploration or minimally invasive PTX instead [9].

Although the preferred sequence of imaging continues to evolve and there exists significant institutional variation, in some institutions, SPS is performed following neck US to confirm the identification of enlarged PTG, or to exclude an ectopic localization before surgical treatment is considered [22,23]. A meta-analysis assessing the value of US imaging in pHPT reported an overall pooled sensitivity of 76.1% and a positive predictive value of 93.2%; however, significant operator and center dependency was emphasized [24]. Nevertheless, despite the wide availability of SPS, several studies have reported great variation in its diagnostic accuracy, with reported sensitivity from 25.4% for US-negative patients to 78.9% for single-gland disease [9,25]. Hence, the quality of both imaging investigations and their interpretation vary dramatically between institutions, and both common localizing tests (US and SPS) may have suboptimal localizing accuracy with a considerable rate of non-localized cases [26].

In the present study, endocrinologists performed all US tests; however, the rate of successful localizations of enlarged PTG was relatively low, and the results of SPS diagnostics were only slightly better. This likely contributed to the failed preoperative imaging of enlarged PTG. Therefore, we implemented 4D-CT scan for preoperative PTG imaging in 2018 at our institution. Owing to this, we achieved a 94.2% success rate of detecting enlarged PTG. Even more importantly, there occurred several favorable changes from the clinical point of view. After the implementation of 4D-CT imaging, the need for redo surgery decreased from 11.5% to 7.3%. We also saw a move towards less extensive and more focused surgical dissection. This was illustrated by shorter PTX operative times; a lower number of patients undergoing the removal of two or more PTG; a lower rate of simultaneous thyroid surgery; and lower morbidity. However, the last difference remained statistically non-significant.

Shorter operative times for PTX in the 4D-CT imaged patients were expected, as the precise preoperative localization of enlarged PTG enabled us to use unilateral focused exploration instead of BNE in most cases. We mostly used a small (4–5 cm) central Kocher's skin incision, followed by a unilateral dissection on the side of the enlarged PTG. The potential benefit of using Kocher's incision is the easy conversion from unilateral dissection to classical BNE in the case of failed one-side exploration. However, according to consensus, PTX is minimally invasive when performed through an incision of less than 3 cm. Therefore, our approach has to be defined as non-minimally invasive [27,28]. Despite this, advances in preoperative imaging certainly promote the use of minimally invasive PTX, with lower complication rates and better cosmetic effect compared to BNE [29].

An additional tool facilitating a more focused dissection and better guidance of surgery was the intraoperative display of the results of 4D-CT PTG imaging. However, this statement is based on professional experience.

We believe that significantly higher rate of extirpation of two or more PTG in the group of preoperatively non-localized PTG was caused by difficulties linked to the intraoperative localization of culprit PTG, especially in cases of small adenomas. The exploration of a slightly enlarged PTG resulted in its extirpation. If the macroscopic changes were 'not convincing', or the frozen section did not confirm the diagnosis of PTG adenoma, further explorative dissection and extirpation of a second PTG, whether with minimal or more 'convincing' changes, was undertaken. As the rate of true histologically proven multiglandular disease for the groups was similar, we could conclude that the use of 4D-CT enables the reduction of the removal rate of 'innocent PTG'.

The uncertain location of PTG probably led to the almost twofold rate of simultaneous thyroid surgery (52.2%) in patients with failed preoperative imaging of the PTG pathology, compared to cases of successfully localized enlarged PTG (28.3–28.6%). The higher rate of concurrent thyroidectomies might have been due to the removal of the protruding thyroid tissue, especially in the retrothyroideal position, which were mistaken for PTG adenomas. Ryan et al. reported a rate of 25% for simultaneous thyroid surgery, and they performed thyroid resection to optimize operative access, ensure the complete removal of involved PTG without the risk of seeding the parathyroid tissue, or more commonly, to remove synchronous thyroid pathology [30]. Thus, the optimal rate for a concurrent thyroid surgery seems to be one in every four cases of PTX. In our study we only saw similar rates for image-positive patients. However, as the rate of simultaneous thyroid surgery reported in other studies is relatively high as well (up to 50%), we believe that this is a place for further research and discussion [25].

After the implementation of 4D-CT, the observed decrease in morbidity was most probably attributable to the less extensive surgical exploration and the discarding of routine BNE. Although the difference in morbidity between the groups was obvious (2.0% for 4D-CT, 13.2% for US or SPS localized, and 10.0% for non-localized cases), it remained on the verge of statistical significance ($p = 0.076$). Nevertheless, we share the opinion that surgeons should not perform blind exploration, as positive imaging investigation can almost always be obtained to make the surgery safer [23].

There are several other diagnostic options that are suggested to improve the results of surgical treatment of pHPT. Positron emission tomography (PET) with 11C-choline has shown outstanding results with an accuracy of almost 99% [31]. However, despite its excellent accuracy, PET has considerable limitations mostly related to its cost and availability [32]. In the present study, intraoperative diagnostic tests (e.g., PTH measurement and frozen section consultation) were used relatively rarely, as they are deemed unacceptable at our institution for being too time consuming. However, we are looking forward to implementing rapid intraoperative PTH assay, as encouraged by a systematic review by Medas et al. reporting a significantly lower rate of persistent and recurrent pHPT with the use of this test [33]. This technique eliminates the need for intraoperative frozen section consultation, thereby saving time and expense [23]. In summary, we think that other novel diagnostic modalities should be used in concurrence with 4D-CT in low-volume centers.

This study had some limitations. First, the retrospective nature of this study means that there were some missing data points. Second, our cohort has quite an impressive female predominance, which can affect the generalizability of our data.

5. Conclusions

The use of 4D-CT imaging enabled to precisely locate almost 95% of enlarged PTG in patients with pHPT. The accurate localization and intraoperatively displayed imaging results are useful guides for surgeons to make PTX a faster and safer procedure in a low-volume center.

Author Contributions: M.M., M.K., K.K., Ü.K., A.T. and K.T. conceptualized the study. M.M. prepared all the documents. M.K. led the data collection process. M.M. drafted the manuscript, assisted by M.K., K.K. and A.T. Ü.K. performed the statistical analysis. All authors have read and agreed to the published version of the manuscript.

Funding: This research received no external funding.

Institutional Review Board Statement: Approval for the study was obtained from the Research Ethics Committee of the University of Tartu (number 300-T4, 20/01/2020).

Informed Consent Statement: Not applicable.

Data Availability Statement: Study data can be made available on a reasonable request for future collaborative analyses.

Conflicts of Interest: The authors declare no conflict of interest.

Abbreviations

BNE	bilateral neck exploration
CT	computed tomography
HPT	hyperparathyroidism
iCa	ionized calcium
MRI	magnetic resonance imaging
pHPT	primary hyperparathyroidism
PET	positron emission tomography
PTG	parathyroid gland
PTH	parathyroid hormone
PTX	parathyroidectomy
RLN	recurrent laryngeal nerve
SPS	^{99m}Tc sestamibi parathyroid scan
US	Ultrasonography
4D-CT	four-dimensional computed tomography

References

1. Minisola, S.; Arnold, A.; Belaya, Z.; Brandi, M.L.; Clarke, B.L.; Hannan, F.M.; Hofbauer, L.C.; Insogna, K.L.; Lacroix, A.; Liberman, U.; et al. Epidemiology, Pathophysiology, and Genetics of Primary Hyperparathyroidism. *J. Bone Miner. Res.* **2022**, *37*, 2315–2329. [CrossRef] [PubMed]
2. Donatini, G.; Marciniak, C.; Lenne, X.; Clément, G.; Bruandet, A.; Sebag, F.; Mirallié, E.; Mathonnet, M.; Brunaud, L.; Lifante, J.C.; et al. Risk Factors of Redo Surgery After Unilateral Focused Parathyroidectomy: Conclusions from a Comprehensive Nationwide Database of 13,247 Interventions Over 6 Years. *Ann. Surg.* **2020**, *272*, 801–806. [CrossRef]
3. Bollerslev, J.; Schalin-Jäntti, C.; Rejnmark, L.; Siggelkow, H.; Morreau, H.; Thakker, R.; Sitges-Serra, A.; Cetani, F.; Marcocci, C. Unmet therapeutic, educational and scientific needs in parathyroid disorders: Consensus statement from the first European Society of Endocrinology Workshop (PARAT). *Eur. J. Endocrinol.* **2019**, *181*, P1–P19. [CrossRef]
4. Clarke, B.L. Asymptomatic primary hyperparathyroidism. *Parathyr. Disord.* **2019**, *51*, 13–22.
5. Wilhelm, S.M.; Wang, T.S.; Ruan, D.T.; Lee, J.A.; Asa, S.L.; Duh, Q.Y.; Doherty, G.M.; Herrera, M.F.; Pasieka, J.L.; Perrier, N.D.; et al. The American association of endocrine surgeons guidelines for definitive management of primary hyperparathyroidism. *JAMA Surg.* **2016**, *151*, 959–968. [CrossRef] [PubMed]
6. Khokar, A.M.; Kuchta, K.M.; Moo-Young, T.A.; Winchester, D.J.; Prinz, R.A. Parathyroidectomy is Safe in Elderly Patients: A National Surgical Quality Improvement Program Study. *World J. Surg.* **2020**, *44*, 526–536. [CrossRef]
7. Zanocco, K.; Heller, M.; Sturgeon, C. Cost-effectiveness of parathyroidectomy for primary hyperparathyroidism. *Endocr. Pract.* **2011**, *17*, 69–74. [CrossRef]
8. Ullmann, T.M.; Adam, M.A.; Sosa, J.A. Surgeon Volume and Outcomes in Primary Hyperparathyroidism—What Is Old Is New Again. *JAMA Surg.* **2022**, *157*, 589. [CrossRef]
9. Naik, M.; Khan, S.R.; Owusu, D.; Alsafi, A.; Palazzo, F.; Jackson, J.E.; Harvey, C.J.; Barwick, T.D. Contemporary Multimodality Imaging of Primary Hyperparathyroidism. *Radiographics* **2022**, *42*, 841–860. [CrossRef]
10. Udelsman, R.; Lin, Z.; Donovan, P. The superiority of minimally invasive parathyroidectomy based on 1650 consecutive patients with primary hyperparathyroidism. *Ann. Surg.* **2011**, *253*, 585–591. [CrossRef]
11. Bilezikian, J.P.; Potts, J.T.; El-Hajj Fuleihan, G.; Kleerekoper, M.; Neer, R.; Peacock, M.; Rastad, J.; Silverberg, S.J.; Udelsman, R.; Wells, S.A. Summary statement from a workshop on asymptomatic primary hyperparathyroidism: A perspective for the 21st century. *J. Clin. Endocrinol. Metab.* **2002**, *87*, 5353–5361. [CrossRef] [PubMed]
12. Bunch, P.M.; Kelly, H.R. Preoperative imaging techniques in primary hyperparathyroidism: A review. *JAMA Otolaryngol. Neck Surg.* **2018**, *144*, 929–937. [CrossRef]
13. Erinjeri, N.J.; Udelsman, R. Volume–outcome relationship in parathyroid surgery. *Best. Pract. Res. Clin. Endocrinol. Metab.* **2019**, *33*, 101287. [CrossRef] [PubMed]
14. Gray, W.K.; Navaratnam, A.V.; Day, J.; Wass, J.A.H.; Briggs, T.W.R.; Lansdown, M. Volume-outcome associations for parathyroid surgery in England: Analysis of an administrative data set for the Getting It Right First Time program. *JAMA Surg.* **2022**, *157*, 581–588. [CrossRef] [PubMed]
15. Sosa, J.A.; Powe, N.R.; Levine, M.A.; Udelsman, R.; Zeiger, M.A. Profile of a clinical practice: Thresholds for surgery and surgical outcomes for patients with primary hyperparathyroidism: A national survey of endocrine surgeons. *J. Clin. Endocrinol. Metab.* **1998**, *83*, 2658–2665. [CrossRef]
16. Nikkolo, C.; Saar, S.; Sokirjanski, M.; Junkin, L.K.; Lepner, U. Retrospektiivne uuring primaarse hüperparatüreoosi kirurgilise ravi tulemustest Tartu Ülikooli Kliinikumis. *Eesti Arst* **2014**, *93*, 622–626.
17. Kettle, A.G.; O'Doherty, M.J. Parathyroid Imaging: How Good Is It and How Should It Be Done? *Semin. Nucl. Med.* **2006**, *36*, 206–211. [CrossRef]

18. Iacobone, M.; Scerrino, G.; Palazzo, F.F. Parathyroid surgery: An evidence-based volume—Outcomes analysis: European Society of Endocrine Surgeons (ESES) positional statement. *Langenbeck's Arch. Surg.* **2019**, *404*, 919–927. [CrossRef]

19. Mitchell, J.; Milas, M.; Barbosa, G.; Sutton, J.; Berber, E.; Siperstein, A. Avoidable reoperations for thyroid and parathyroid surgery: Effect of hospital volume. *Surgery* **2008**, *144*, 899–907. [CrossRef]

20. Frank, E.; Watson, W.A.; Fujimoto, S.; De Andrade Filho, P.; Inman, J.; Simental, A. Surgery versus Imaging in Non-Localizing Primary Hyperparathyroidism: A Cost-Effectiveness Model. *Laryngoscope* **2020**, *130*, E963–E969. [CrossRef]

21. Minisola, S.; Cipriani, C.; Diacinti, D.; Tartaglia, F.; Scillitani, A.; Pepe, J.; Scott-Coombes, D. Imaging of the parathyroid glands in primary hyperparathyroidism. *Eur. J. Endocrinol.* **2016**, *174*, D1–D8. [CrossRef] [PubMed]

22. Morris, M.A.; Saboury, B.; Ahlman, M.; Malayeri, A.A.; Jones, E.C.; Chen, C.C.; Millo, C. Parathyroid Imaging: Past, Present, and Future. *Front. Endocrinol.* **2022**, *12*, 760419. [CrossRef] [PubMed]

23. Udelsman, R. Approach to the patient with persistent or recurrent primary hyperparathyroidism. *J. Clin. Endocrinol. Metab.* **2011**, *96*, 2950–2958. [CrossRef]

24. Cheung, K.; Wang, T.S.; Farrokhyar, F.; Roman, S.A.; Sosa, J.A. A meta-analysis of preoperative localization techniques for patients with primary hyperparathyroidism. *Ann. Surg. Oncol.* **2012**, *19*, 577–583. [CrossRef]

25. Lenschow, C.; Wennmann, A.; Hendricks, A.; Germer, C.T.; Fassnacht, M.; Buck, A.; Werner, R.A.; Plassmeier, L.; Schlegel, N. Questionable value of [99mTc]-sestamibi scintigraphy in patients with pHPT and negative ultrasound. *Langenbeck's Arch. Surg.* **2022**, *407*, 3661–3669. [CrossRef]

26. Pappu, S.; Donovan, P.; Cheng, D.; Udelsman, R. Sestamibi scans are not all created equally. *Arch. Surg.* **2005**, *140*, 383–386. [CrossRef]

27. Brunaud, L.; Zarnegar, R.; Wada, N.; Ituarte, P.; Clark, O.H.; Duh, Q.Y. Incision length for standard thyroidectomy and parathyroidectomy: When is it minimally invasive? *Arch. Surg.* **2003**, *138*, 1140–1143. [CrossRef] [PubMed]

28. Bellantone, R.; Raffaelli, M.; De Crea, C.; Traini, E.; Lombardi, C.P. Minimally-invasive parathyroid surgery. *Acta Otorhinolaryngol. Ital.* **2011**, *31*, 207–215.

29. Ishii, H.; Mihai, R.; Watkinson, J.C.; Kim, D.S. Systematic review of cure and recurrence rates following minimally invasive parathyroidectomy. *BJS Open.* **2018**, *2*, 364–370. [CrossRef]

30. Ryan, S.; Courtney, D.; Moriariu, J.; Timon, C. Surgical management of primary hyperparathyroidism. *Eur. Arch. Oto-Rhino-Laryngol.* **2017**, *274*, 4225–4232. [CrossRef]

31. Orevi, M.; Freedman, N.; Mishani, E.; Bocher, M.; Jacobson, O.; Krausz, Y. Localization of parathyroid adenoma by 11C-choline PET/CT: Preliminary results. *Clin. Nucl. Med.* **2014**, *39*, 1033–1038. [CrossRef] [PubMed]

32. Tay, D.; Das, J.P.; Yeh, R. Preoperative localization for primary hyperparathyroidism: A clinical review. *Biomedicines* **2021**, *9*, 390. [CrossRef] [PubMed]

33. Medas, F.; Cappellacci, F.; Canu, G.L.; Noordzij, J.P.; Erdas, E.; Calò, P.G. The role of Rapid Intraoperative Parathyroid Hormone (ioPTH) assay in determining outcome of parathyroidectomy in primary hyperparathyroidism: A systematic review and meta-analysis. *Int. J. Surg.* **2021**, *92*, 106042. [CrossRef] [PubMed]

Article

Sporadic Parathyroid Adenoma: A Pilot Study of Novel Biomarkers in Females

Angeliki Cheva [1,†], Angeliki Chorti [2,*,†], Kassiani Boulogeorgou [1], Anthoula Chatzikyriakidou [3], Charoula Achilla [3], Vangelis Bontinis [4], Alkis Bontinis [4], Stefanos Milias [5], Thomas Zarampoukas [6], Sohail Y. Bakkar [7] and Theodosios Papavramidis [2,5]

[1] Laboratory of Pathology, Faculty of Health Science, Medical School, Aristotle University, 541 24 Thessaloniki, Greece
[2] 1st Propaedeutic Department of Surgery, Faculty of Health Science, Medical School, AHEPA University Hospital, Aristotle University, 546 36 Thessaloniki, Greece
[3] Laboratory of Medical Biology—Genetics, Faculty of Health Science, Medical School, Aristotle University, 546 36 Thessaloniki, Greece
[4] Department of Vascular Surgery, Faculty of Health Science, Medical School, AHEPA University Hospital, Aristotle University, 546 36 Thessaloniki, Greece
[5] Minimal Invasive Endocrine Surgery Department, Kyanos Stavros, Euromedica, 546 36 Thessaloniki, Greece
[6] Laboratory of Pathology, Interbalkan Medical Center, 546 26 Thessaloniki, Greece
[7] Endocrine & General Surgery, The Hashemite University, Amman 13133, Jordan
* Correspondence: angeliki.g.chorti@gmail.com
† These authors contributed equally to this work.

Abstract: *Background and Objectives*: Parathyroid adenoma is a distinct cause of primary hyperparathyroidism, with the vast majority being sporadic ones. Proteomic analysis of parathyroid adenomas has proposed a large number of related proteins. The aim of this study is to evaluate the immunohistochemical staining of ANXA2, MED12, MAPK1 and VDR in parathyroid adenoma tissue. *Materials and Methods*: Fifty-one parathyroid adenomas were analyzed for ANXA2, MED12, MAPK1 and VDR expressions. Tissue was extracted from formalin-fixed paraffin-embedded parathyroid adenoma specimens; an immunohistochemical study was applied, and the percentage of allocation and intensity were evaluated. *Results*: ANXA2 stained positively in 60.8% of all cell types, while MED12 had positive staining in 66%. MAPK1 expression was found to be negative in total, although a specific pattern for oxyphil cells was observed, as they stained positive in 17.7%. Finally, VDR staining was positive at 22.8%, based on nuclear staining. *Conclusions*: These immunohistochemical results could be utilized as biomarkers for the diagnosis of sporadic parathyroid adenoma. It is of great importance that a distinct immunophenotype of nodule-forming cells in a positive adenoma could suggest a specific pattern of adenoma development, as in hereditary patterns.

Keywords: immunohistochemistry; parathyroid adenoma; ANXA2; MED12; MAPK1; VDR

Citation: Cheva, A.; Chorti, A.; Boulogeorgou, K.; Chatzikyriakidou, A.; Achilla, C.; Bontinis, V.; Bontinis, A.; Milias, S.; Zarampoukas, T.; Bakkar, S.Y.; et al. Sporadic Parathyroid Adenoma: A Pilot Study of Novel Biomarkers in Females. *Medicina* 2024, 60, 1100. https://doi.org/10.3390/medicina60071100

Academic Editor: Jin Wook Yi

Received: 29 May 2024
Revised: 28 June 2024
Accepted: 3 July 2024
Published: 5 July 2024

1. Introduction

Primary hyperparathyroidism (PHPT) affects approximately 1% of the population and is characterized by hypersecretion of parathyroid hormone (PTH) and serum-ionized calcium. There is a female prevalence, with a ratio between men and women of 3–4:1. There are three distinct subtypes of this disease: the most commonly found single parathyroid adenoma (80–85%), multi-glandular hyperplasia (10–15%) and parathyroid carcinoma (1%) [1–4].

Parathyroid adenomas are almost always (90%) sporadic, with the exception of MEN (Multiple Endocrine Neoplasia) syndrome and the HPT-jaw tumor (hyperparathyroidism-jaw tumor) syndrome. Non-syndromic hereditary forms of PHPT, such as FIHPT (familiar hyperparathyroidism), FHH (familial hypocalciuric hypercalcemia) and NS-HPT (neonatal severe hyperparathyroidism) are other causes of PHPT [5].

In the international literature, proteomic analysis of parathyroid adenoma tissues has been reported, in which Western blotting, MALDI/TOF (matrix-assisted laser desorption/ionization), mass spectrometry and immunohistochemistry were applied. A vast number of proteins were found to have altered expression in parathyroid adenomas [6–11].

Annexin-2 (ANXA2) is a 36-kDa protein with NH2—terminal binding ionized calcium, a C-terminal binding actin-F and phosphoinositol, and a central nucleus that creates stable a-helical disks [12,13]. ANXA2 controls cell development and proliferation as well as cell death by interacting with p53 and Nf-kB. Furthermore, ANXA2 regulates cell division, promoting changes in the cytoskeleton, and has been found in increased quantities in the G1-S phase of the cell cycle. It participates in cytoskeleton stabilization and cell interactive activity, as it promotes cell signaling bound with ionized calcium [13]. ANXA2 plays an important role in distinct disease pathogenesis, such as osteoporosis, osteonecrosis in sickle cell anemia and depression, as well as in carcinogenesis, involved in neovascularization and in the development, migration and metastasis of cancer cells [12–17].

Mediator complex subunit 12 (MED12) constitutes one part of the complex that initiates DNA transcription by binding RNA polymerase II with specific transcription factors and interacting with specific regulatory proteins [18]. It is located in the X-chromosome and is structured by 2212 amino acids and is separated into four distinct parts from the N- to the C-terminal end: leucine part, leucine–serine part, proline–glutamine–leucine part and Opa part [19]. MED12 has been correlated with multiple syndromes such as FG, Lujan, X-linked Ohdo, Opitz–Kaveggia syndrome, as well as neuropsychiatric disorders and tachyarrythmias [20–22]. Furthermore, it has been linked to uterine leiomyomas, breast adenomas and phylloid tumors [23–25].

Mitogen-activated protein kinase 1 (MAPK1) constitutes a part of the MAPK/Erk signaling pathway. MAPK1 is a 41kDA protein that is involved in mitosis, cell proliferation and death [26–28]. Its role in carcinogenesis through epithelial–mesenchymal transition has been extensively studied, while MAPK1 has evolved in hypertrophic cardiomyopathy, coronary artery disease and schizophrenia [28].

Vitamin D receptor (VDR) is a 427 amino acids protein with an N- and C-terminal and a DNA-binding area [29]. VDR acts as a calcitriol mediator in the nucleus, where it forms a heterodimer with retinoid X receptor. VDR is involved in thyroid follicular, colorectal, salivary gland and adrenal gland adenomas [30–33]. In parathyroid adenoma, it has already been found that VDR expression is decreased and thus in oxyphil cells [6,9].

The aim of our study is to evaluate the quantitative expression of four gene proteins (ANXA2, MED12, MAPK1 and VDR) by immunohistochemical analysis in tissue samples of parathyroid adenoma, as well as their qualitative characteristics.

2. Materials and Methods

2.1. Tissue Preparation and Assessment

Fifty-nine patients who underwent surgical excision of a single parathyroid adenoma from 2019 to 2022 were enrolled in this study. The diagnosis of primary hyperparathyroidism caused by sporadic parathyroid adenoma was made by endocrinological investigation prior to surgical treatment [34]. Patients with a family history of parathyroid adenoma or Multiple Endocrine Neoplasia Type II syndrome were excluded from our study. All patients provided their consent for participation in the study, which was approved by the local ethics committee. Serum-ionized calcium, phosphorus and parathormone levels were evaluated preoperatively (on the day of the operation) and on the 1st postoperative day.

Specimens' formalin-fixed paraffin-embedded (FFPE) blocks were obtained and assessed, and 9 patients were excluded due to insufficient material. From the final total of 50 FFPE blocks, 2 μm sections were taken, stained with Hematoxylin–Eosin (H&E), as well as sections on specific positive charged slides applied for immunohistochemical stains.

2.2. Immunohistochemical Staining of ANXA2, MED12, MAPK1 and VDR

Afterward, the immunohistochemical staining of ANXA2, MED12, MAPK1 and VDR proteins in parathyroid adenoma was assessed. These specific proteins were chosen according to their association with the formation of other adenomas, either endocrine or not, and the immunohistochemical results were cross-examined with genotyping of specific polymorphisms of these genes so that their association with sporadic parathyroid adenoma to be certified. The antibodies applied were as follows: (a) ANXA2, monoclonal (clone C-10) (from SANTA CRUZ, USA) in a dilution of 1:1000 (45 min) (microwave pH = 6 for 16 min); (b) MED12, rabbit polyclonal (from ATLAS/SIGMA, Sweden) in dilution 1:20 (120 min) (microwave pH = 9 for 20 min); (c) MAPK1, rabbit polyclonal (from ATLAS/SIGMA, Sweden) at a dilution of 1:50 (60 min) (microwave pH = 6 for 16min); (d) VDR, rabbit polyclonal (from SANTA CRUZ USA) at a 1:50 dilution (60 min) (microwave pH = 9 for 20 min). Immunohistochemistry was performed on 2 μm sections on a DAKO Autostainer Link automated platform (DAKO Agilent, USA) with EnVision Flex DAKO detection system (DAKO Agilent, USA). The sections were incubated at 60°C for 16–18 h, deparaffinized and dehydrated. Afterward, antigen retrieval and incubation with each monoclonal antibody were applied. The evaluation was performed in an optical microscope with 20× and 40× magnification.

Across every immunohistochemical staining, external controls were incorporated. These controls encompassed two essential components: (a) specimens known to elicit a positive reaction to the targeted proteins and (b) specimens exhibiting a negative reaction. This procedure was undertaken to ensure a robust and reliable assessment of the outcome.

For the ANXA2 antibody, lung parenchyma served as the designated external positive control. Similarly, brain substance cells were utilized as the external positive control for the MAPK1 antibody, and glandular epithelial cells of the colon were employed for the MED12 antibody. Lastly, Hodgkin cells from classical Hodgkin lymphoma of the nodular sclerosis type, along with brain cells, were utilized as external controls for the VDR antibody.

Moreover, within each slide of immunohistochemical staining, the intensity of the positive reaction was juxtaposed against endogenous controls, namely cells inherently expressing the specific proteins under examination. In the case of ANXA2 and MAPK1 antibodies, the vascular endothelium was designated as the internal positive control. Conversely, for the VDR antibody, the internal positive control consisted of a hypopopulation of T-lymphocytes (FOXP3+) and sporadic positive nuclei within the normal parathyroid gland, while the vascular endothelium was deemed the internal negative control. The surrounding normal parathyroid tissue was also evaluated for its immunohistochemical reaction with these antibodies.

Staining was scored as follows:

(A) Allocation. The percentage of each type of parathyroid gland cell involved in the adenoma (over all adenoma cells) that was positive for each staining was assessed and scored as follows: i. 0: 0–5% rate; ii. 1: rate 6–30%; iii. 2: 31–70% rate; iv. 3: rate 71–100%.

(B) Intensity. Graded as negative, mild, moderate, intense compared to the internal control (vessels, adipose tissue, normal parathyroid gland): i. 0: negative; ii. 1: soft; iii. 2: moderate; iv. 3: intense.

(C) Type of staining. Staining was scored as cytoplasmic, membranous or nuclear depending on the part of the cell targeted: i. K: cytoplasmic; ii. M: membranous; iii. P: nuclear.

The assessment was carried out by two independent pathologists, and adenomas were classified as positive for allocation and intensity categories 1–3, i.e., >5% allocation and mild to intense intensity.

2.3. Statistical Analysis

SPSS statistical package 22.0 (SPSS Inc., Chicago, IL, USA) was used to test frequencies of immunohistochemical staining. One-way ANOVA and the Kruskal–Wallis test were applied to correlate categorical and numeric variables based on the distribution model. For continuous data, we measured means and standard deviations for normal distribution and medians and variances in case of non-normal distribution. A difference at $p \leq 0.05$ was considered statistically significant.

3. Results

A total number of 50 female patients with a mean age of 54.11 ± 12.46 years old were enrolled in this study. The mean diameter of the adenoma was 1.87 ± 0.73 cm, and the mean weight was 1.12 ± 1.04 gr. Table 1 summarizes patients' characteristics, and Table 2 displays immunohistochemical results.

Table 1. Sporadic adenoma characteristics.

Gender	
Female	50
Age (years)	54.11 ± 12.46
Preoperative serum calcium (mg/dL) (normal values 8.4–10.2)	10.91 ± 0.76
Postoperative serum calcium (mg/dL) (normal values 8.4–10.2)	9.12 ± 0.62
Preoperative parathyroid hormone (pg/mL) (normal values 10–68)	82.77 ± 71.04
Postoperative parathyroid hormone (pg/mL) (normal values 10–68)	18.11 ± 16.99
Preoperative serum phosphorus (mg/dL) (normal values 2.3–4.7)	2.88 ± 0.46
Postoperative serum phosphorus (mg/dL) (normal values 2.3–4.7)	3.38 ± 0.74
Adenoma diameter (cm)	1.87 ± 0.73
Adenoma weight (gr)	1.12 ± 1.04

The percentage of chief, oxyphil and clear cells in H&E staining were determined: (a) 12 chief cell adenomas; (b) 3 oxyphil cell adenomas; (c) 18 chief and clear cell adenomas; (d) 12 chief and oxyphil cell adenomas; and (e) 5 chief, clear and oxyphil cell adenomas. Then, the total percentage of immunohistochemical staining and the positive result for each cell type were assessed.

3.1. ANXA2 Immunohistochemical Staining

The total percentage of immunohistochemically positive cells with ANXA2 antibody in parathyroid adenoma is 60.8%, which makes ANXA2 positive staining in sporadic parathyroid adenoma (Figure 1).

A total of 51% of chief cells were positive in ANXA2 immunohistochemical staining, with 19.6% showing intense staining, while 46.5% had mild and moderate intensity. The majority of positive chief cells showed mild staining (27.5%). The type of staining in 59.4% was cytoplasmic, 3.9% was membranous, and 21.6% was cytoplasmic and membranous.

For clear cells, the percentage of >5% positive cells was 21.6%, meaning that the majority of cells have negative staining. The intensity of staining is mostly mild, and its type is cytoplasmic.

The majority of oxyphil cells are negative for ANXA staining (62.7%). However, 27.5% of oxyphil cells are positive at a rate of >70%. Staining is mild and cytoplasmic.

In some cases, in the same adenoma, there is a clearly distinct immunophenotype of chief, clear and oxyphil cells.

ANXA2 positive staining was correlated with postoperative serum ionized calcium levels. Thus, when the staining was intense, the postoperative serum ionized calcium levels were higher than in moderate staining ($p = 0.03$).

Table 2. Immunohistochemical results (NA = not applicable).

	Total Positivity	Chief Cells			Oxyphil Cells			Clear Cells		
		Positivity	Intensity	Type of Staining	Positivity	Intensity	Type of Staining	Positivity	Intensity	Type of Staining
ANXA2	60.8%	51%	27.5% mild 19.6% intense 46.5% mild and moderate	59.4% cytoplasmic 3.9% membranous 21.6% cytoplasmic and membranous	37.3%	Mild	cytoplasmic	21.6%	Mild	Cytoplasmic
MED12	66%	86.3%	23.5% mild 23.5% moderate 25.5% mild and moderate 27.5% intense	Nuclear	35.3%	19.6% moderate	Nuclear	33.3%	15.7% moderate	Nuclear
MAPK1	11.8%	Negative	Negative	Negative	17.7%	Mild	Cytoplasmic	Negative	Negative	Negative
VDR (all types)	78.5%	NA	NA	All	NA	NA	All	NA	NA	All
VDR (nuclear type)	22.8%	NA	NA	Nuclear	NA	NA	Nuclear	NA	NA	Nuclear

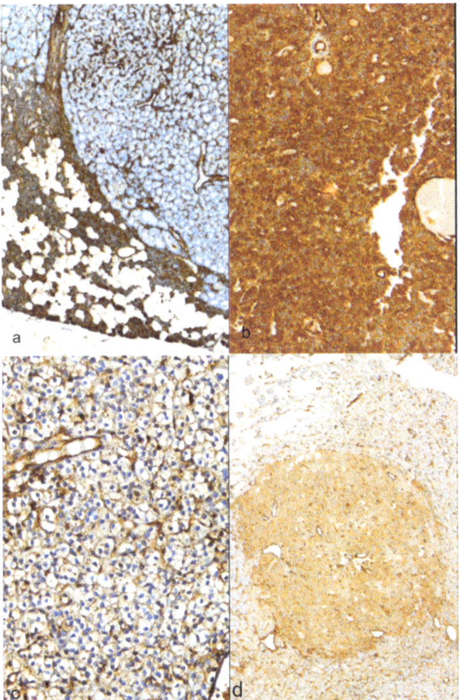

Figure 1. ANXA2 immunohistochemical staining: (**a**) negative staining of chief cells and positive staining of normal parathyroid tissue; (**b**) positive staining of chief cells; (**c**) positive staining of clear cells; (**d**) a distinct immunophenotype of chief cell nodule in an adenoma.

3.2. MED12 Immunohistochemical Staining

Immunohistochemical staining for MED12 is nuclear. The overall rate of positive immunohistochemical staining for MED12 is 66%, demonstrating a positive reaction in the sporadic parathyroid adenoma (Figure 2).

In chief cells, 86.3% have a positive reaction, with 58.8% in the 71–100% positive allocation. The intensity was intense at 27.5%, while a total larger percentage had mild and moderate staining (mild = 23.5%, moderate = 23.5% and mild + moderate = 25.5%).

Regarding clear cells, 33.3% had positive staining, and thus, the positive cells were >70% of the total number of cells. The remaining percentage of cells was negative. The intensity of staining in the majority of cells was moderate at 15.7%.

The majority of oxyphil cells had negative results, with 35.3% of cells having a positive reaction. The intensity of staining was moderate in 19.6%, while mild and intense staining was observed in <10%.

3.3. MAPK1 Immunohistochemical Staining

MAPK1 protein seems to follow a different expression pattern as in our study, it was detected with positive expression mainly in the oxyphil cells of the parathyroid gland.

In total, MAPK1 was negative (88.2% of cells are <5% positive).

Nevertheless, a trend for oxyphil cells was observed, as oxyphil cells were positive, making it a remarkable observation. Oxyphil cells were positive in 17.7% with mild intensity and cytoplasmic color (Figure 3).

MAPK1 positive staining was correlated with adenoma weight. Thus, in moderate staining, the adenoma was weighted more than in cases of negative staining ($p = 0.04$).

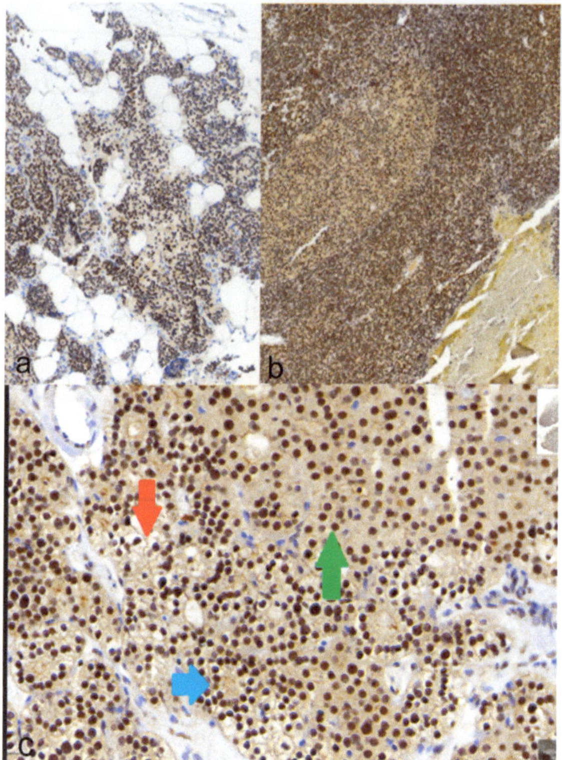

Figure 2. MED12 immunohistochemical staining: (**a**) positivity in normal parathyroid tissue, (**b**) a distinct immunophenotype of oxyphil cell nodule in a positive chief cell adenoma; (**c**) positive chief cells (blue arrow), clear cells (red arrow) and oxyphil cells (green arrow).

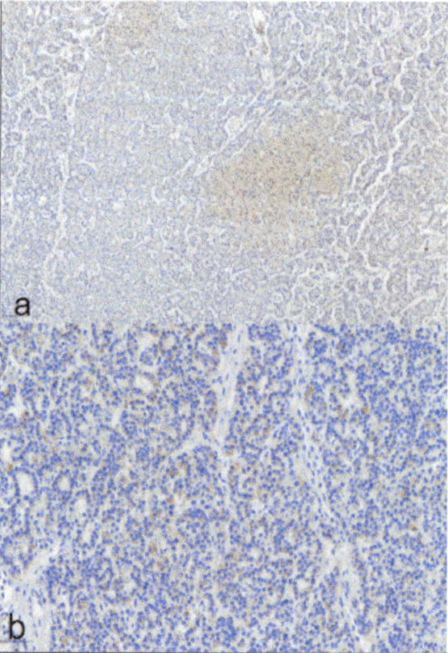

Figure 3. MAPK1 immunohistochemical staining: (**a**) positive oxyphil cells and negative chief cells; (**b**) negative chief cells.

3.4. VDR Immunohistochemical Staining

According to the existing literature, VDR staining is nuclear [35,36].

The mean value for positive results based on nuclear staining is 22.8% (Figure 4).

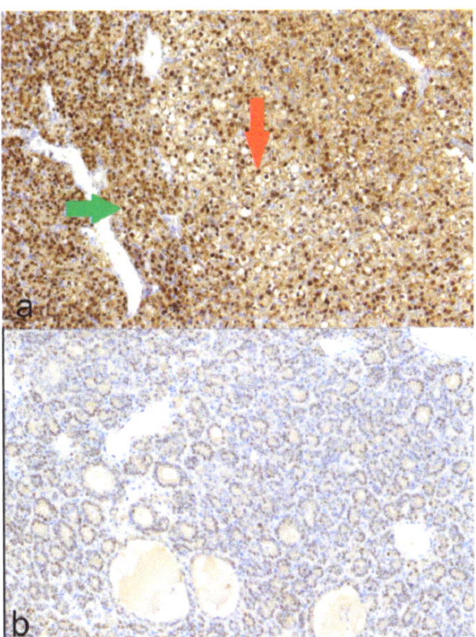

Figure 4. VDR immunohistochemical staining: (**a**) positive nuclear staining in chief (green arrow) and clear (red arrow) cells; (**b**) negative (blue) and positive (brown) nuclear staining in chief cells.

Finally, a clearly distinct immunophenotype was observed in the same adenoma, with groups of nodule-forming cells being either negative in staining or mild intensity compared to the rest of the adenoma that had an intense reaction (Figure 5).

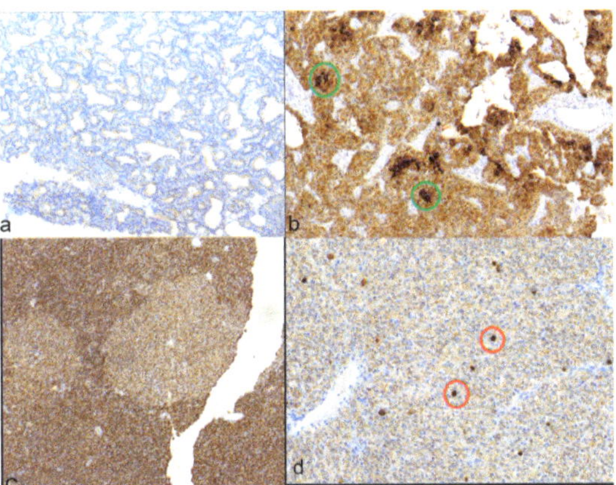

Figure 5. VDR immunohistochemical staining: (**a**) positive membranous staining in the apical side of chief cells; (**b**) positive, intense cytoplasmic staining in group of chief cells (green circle) encircled by positive moderate cytoplasmic staining; (**c**) a distinct immunophenotype of chief cell nodule with mild intensity in an intense positive chief cell adenoma; (**d**) intense positive membranous staining in group of chief cells (red circle) encircled by mild positive membranous staining chief cells.

4. Discussion

Immunohistochemical analysis of ANXA2, MED12, MAPK1 and VDR proteins in sporadic parathyroid adenoma confirmed their expression in parathyroid cells. The ANXA2 protein appears to be expressed in all three cell types (main, clear and oxyphil) of parathyroid cells, and immunohistochemical testing revealed a positive staining reaction in adenomas. This result seems to be in line with similar studies in the international literature. Giusti et al. studied the proteins expressed in the normal parathyroid gland compared to the adenoma using mass spectrometry and identified a number of proteins whose expression is altered in the adenoma, including ANXA2, which is triplicated and associated with cell apoptosis [11]. These results were also confirmed by the study by Akpinar et al. with the same methodology, while immunohistochemistry is recommended to confirm the results in these studies [8,10,11].

Furthermore, MAPK1 was found to have weak expression in oxyphil adenomas. The different expression between normal tissue and adenomas was shown to be statistically significant [11]. The MAPK1 signaling pathway seems to play a crucial role in the pathogenesis of parathyroid adenomas and includes a number of other proteins with different expressions in adenomas [7,9,11]. The role of MAPK1 in cell signal transduction seems to be decisive in the pathogenesis of the disease, while the increased expression of mitochondrial proteins in adenomas seems to be related to the conversion of chief cells to oxyphil ones in adenomas and possibly explains the results of our own study [10].

In our study, VDR appeared to be expressed in parathyroid adenoma. The results of the study comparing gene expression in chief and oxyphil cell adenomas by Lu et al. reported reduced expression levels (mild-moderate intensity) of VDR protein in both types of adenomas compared to normal tissue (strong expression) [6]. The reduced intensity of expression of this protein in adenomas was also observed in a previous study by Rao et al. in 2000 [6,37]. This modification is probably related to the increased values of calcium and parathormone in serum (6). In our study, the type of VDR immunohistochemical staining was assessed in two arms: (a) nuclear staining and (b) cytoplasmic and membranous staining. In the international literature, there is a disconcordance of the type of staining for this specific antibody that is considered specific for evaluating the positive reaction in the cells. In the study by An et al., cytoplasmic and nuclear staining is reported in endometrioid carcinoma cells, while in the study by Rehab Mohamed Sharaf et al., cytoplasmic and membrane staining of the VDR antibody in urothelial carcinoma cells is reported [38,39].

The expression of MED12 is reported to be altered in parathyroid gland adenomas in previous studies with mass spectrometry [7]. In our study, MED12 protein expression was found to be positive in the majority of cells, especially in chief cells. Arya et al. reported the expression of this protein in adenoma cells after mass spectrometry [7].

A limitation of our study is the small sample of specimens, as it is formed as a pilot study, and further analysis will be conducted. Another limitation is the inclusion only of women in our study, as primary hyperparathyroidism affects predominantly females. Furthermore, our study does not have a cross-sectional analysis, which could be a possible limitation of our study, but a cross-sectional study is already ongoing.

5. Conclusions

ANXA2, MED12, MAPK1 and VDR proved to have a positive staining in the immunohistochemical study of sporadic parathyroid adenomas in varying intensity and allocation percentages. These findings enable the application of these proteins as biomarkers for the diagnosis of parathyroid adenoma. In addition, this study demonstrates that in sporadic parathyroid adenoma, as in hereditary syndromes, there are clearly distinct immunophenotypes of some types of nodule-forming cells in the same adenoma, which probably describes a specific pattern of adenoma development.

Author Contributions: Conceptualization, A.C. (Angeliki Cheva) and A.C. (Angeliki Chorti); methodology, A.C. (Angeliki Cheva), A.C. (Angeliki Chorti) and K.B.; software, K.B.; validation, A.C. (Angeliki Chorti), A.C. (Angeliki Cheva), K.B. and A.C. (Anthoula Chatzikyriakidou); formal analysis, A.C. (Angeliki Chorti) and A.C. (Angeliki Cheva); investigation, A.C. (Angeliki Chorti), V.B., A.B., C.A., S.M. and T.Z.; resources, A.C. (Angeliki Chorti); data curation, A.C. (Angeliki Chorti), V.B., A.B., C.A., S.M. and T.Z.; writing—original draft preparation, A.C. (Angeliki Chorti) and A.C. (Angeliki Cheva); writing—review and editing, T.P. and S.Y.B.; supervision, A.C. (Angeliki Cheva) and T.P.; project administration, T.P. All authors have read and agreed to the published version of the manuscript.

Funding: This research received no external funding.

Institutional Review Board Statement: The study was conducted in accordance with the Declaration of Helsinki and approved by the Institutional Review Board (or Ethics Committee) of Aristotle University of Thessaloniki (1300/20 October 2020).

Informed Consent Statement: Informed consent was obtained from all subjects involved in the study.

Data Availability Statement: Data will be available upon reasonable request.

Conflicts of Interest: The authors declare no conflicts of interest.

References

1. Bilezikian, J.P.; Brandi, M.L.; Eastell, R.; Silverberg, S.J.; Udelsman, R.; Marcocci, C.; Potts, J.T. Guidelines for the management of asymptomatic primary hyperparathyroidism: Summary statement from the Fourth International Workshop. *J. Clin. Endocrinol. Metab.* **2014**, *99*, 3561–3569. [CrossRef] [PubMed]
2. Cusano, N.E.; Cetani, F. Normocalcemic primary hyperparathyroidism. *Arch. Endocrinol. Metab.* **2022**, *66*, 666–677. [CrossRef] [PubMed]
3. Zavatta, G.; Clarke, B.L. Normocalcemic Hyperparathyroidism: A Heterogeneous Disorder Often Misdiagnosed? *JBMR Plus* **2020**, *4*, e10391. [CrossRef] [PubMed]
4. Erickson, L.A.; Mete, O.; Juhlin, C.C.; Perren, A.; Gill, A.J. Overview of the 2022 WHO Classification of Parathyroid Tumors. *Endocr. Pathol.* **2022**, *33*, 64–89. [CrossRef] [PubMed]
5. Cetani, F.; Pardi, E.; Borsari, S.; Marcocci, C. Molecular pathogenesis of primary hyperparathyroidism. *J. Endocrinol. Invest.* **2011**, *34*, 35–39. [PubMed]
6. Lu, M.; Kjellin, H.; Fotouhi, O.; Lee, L.; Nilsson, I.L.; Haglund, F.; Höög, A.; Lehtiö, J.; Larsson, C. Molecular profiles of oxyphilic and chief cell parathyroid adenoma. *Mol. Cell Endocrinol.* **2018**, *470*, 84–95. [CrossRef] [PubMed]
7. Arya, A.K.; Bhadada, S.K.; Singh, P.; Dahiya, D.; Kaur, G.; Sharma, S.; Saikia, U.N.; Behera, A.; Rao, S.D.; Bhasin, M. Quantitative proteomics analysis of sporadic parathyroid adenoma tissue samples. *J. Endocrinol. Invest.* **2019**, *42*, 577–590. [CrossRef]
8. Donadio, E.; Giusti, L.; Cetani, F.; Da Valle, Y.; Ciregia, F.; Giannaccini, G.; Pardi, E.; Saponaro, F.; Torregrossa, L.; Basolo, F.; et al. Evaluation of formalin-fixed paraffin-embedded tissues in the proteomic analysis of parathyroid glands. *Proteome Sci.* **2011**, *9*, 29. [CrossRef] [PubMed]
9. Varshney, S.; Bhadada, S.K.; Saikia, U.N.; Sachdeva, N.; Behera, A.; Arya, A.K.; Sharma, S.; Bhansali, A.; Mithal, A.; Rao, S.D. Simultaneous expression analysis of vitamin D receptor, calcium-sensing receptor, cyclin D1, and PTH in symptomatic primary hyperparathyroidism in Asian Indians. *Eur. J. Endocrinol.* **2013**, *169*, 109–116. [CrossRef]
10. Akpinar, G.; Kasap, M.; Canturk, N.Z.; Zulfigarova, M.; Islek, E.E.; Guler, S.A.; Simsek, T.; Canturk, Z. Proteomics analysis of tissue samples reveals changes in mitochondrial protein levels in parathyroid hyperplasia over adenoma. *Cancer Genom. Proteom.* **2017**, *14*, 197–211. [CrossRef]
11. Giusti, L.; Cetani, F.; Ciregia, F.; Da Valle, Y.; Donadio, E.; Giannaccini, G.; Banti, C.; Pardi, E.; Saponaro, F.; Basolo, F.; et al. A proteomic approach to study parathyroid glands. *Mol. Biosyst.* **2011**, *7*, 687–699. [CrossRef] [PubMed]
12. Christensen, M.V.; Høgdall, C.K.; Umsen, K.M.J.; Høgdall, E.V.S. Annexin A2 and cancer: A systematic review. *Int. J. Oncol.* **2018**, *52*, 5–18. [CrossRef] [PubMed]
13. Sharma, M.C. Annexin A2 (ANX A2): An emerging biomarker and potential therapeutic target for aggressive cancers. *Int. J. Cancer* **2019**, *144*, 2074–2081. [CrossRef] [PubMed]
14. Deng, F.Y.; Lei, S.F.; Zhang, Y.; Zhang, Y.L.; Zheng, Y.P.; Zhang, L.S.; Pan, R.; Wang, L.; Tian, Q.; Shen, H.; et al. Peripheral blood monocyte-expressed ANXA2 gene is involved in pathogenesis of osteoporosis in humans. *Mol. Cell. Proteom.* **2011**, *10*, M111.011700. [CrossRef] [PubMed]
15. Zhang, X.; Liu, S.; Guo, C.; Zong, J.; Sun, M.Z. The association of annexin A2 and cancers. *Clin. Transl. Oncol.* **2012**, *14*, 634–640. [CrossRef] [PubMed]
16. Jin, J.; Bhatti, D.L.; Lee, K.W.; Medrihan, L.; Cheng, J.; Wei, J.; Zhong, P.; Yan, Z.; Kooiker, C.; Song, C.; et al. Ahnak scaffolds p11/Anxa2 complex and L-type voltage-gated calcium channel and modulates depressive behavior. *Mol. Psychiatry* **2020**, *25*, 1035–1049. [CrossRef]

17. Pandey, S.; Ranjan, R.; Pandey, S.; Mishra, R.M.; Seth, T.; Saxena, R. Effect of ANXA2 gene single nucleotide polymorphism (SNP) on the development of osteonecrosis in Indian sickle cell patient: A PCR-RFLP approach. *Indian. J. Exp. Biol.* **2012**, *50*, 455–458.
18. Available online: https://www.uniprot.org/uniprotkb/Q93074/entry (accessed on 2 July 2024).
19. Wang, H.; Shen, Q.; Ye, L.-H.; Ye, J. MED12 mutations in human diseases. *Protein Cell* **2013**, *4*, 643–646. [CrossRef] [PubMed]
20. Narayanan, D.L.; Phadke, S.R. A novel variant in MED12 gene: Further delineation of phenotype. *Am. J. Med. Genet. A* **2017**, *173*, 2257–2260. [CrossRef]
21. Philibert, R.A.; Madan, A. Role of MED12 in transcription and human behavior. *Pharmacogenomics* **2007**, *8*, 909–916. [CrossRef]
22. Napoli, C.; Schiano, C.; Soricelli, A. Increasing evidence of pathogenic role of the Mediator (MED) complex in the development of cardiovascular diseases. *Biochimie* **2019**, *165*, 1–8. [CrossRef] [PubMed]
23. Pérot, G.; Croce, S.; Ribeiro, A.; Lagarde, P.; Velasco, V.; Neuville, A.; Coindre, J.-M.; Stoeckle, E.; Floquet, A.; MacGrogan, G.; et al. MED12 Alterations in Both Human Benign and Malignant Uterine Soft Tissue Tumors. *PLoS ONE* **2012**, *7*, e40015. [CrossRef] [PubMed]
24. Laé, M.; Gardrat, S.; Rondeau, S.; Richardot, C.; Caly, M.; Chemlali, W.; Vacher, S.; Couturier, J.; Mariani, O.; Terrier, P.; et al. MED12 mutations in breast phyllodes tumors: Evidence of temporal tumoral heterogeneity and identification of associated critical signaling pathways. *Oncotarget* **2016**, *7*, 84428–84438. [CrossRef] [PubMed]
25. Loke, B.N.; Md Nasir, N.D.; Thike, A.A.; Lee, J.Y.H.; Lee, C.S.; Teh, B.T.; Tan, P.H. Genetics and genomics of breast fibroadenomas. *J. Clin. Pathol.* **2018**, *71*, 381–387. [CrossRef] [PubMed]
26. Available online: https://atlasgeneticsoncology.org/gene/41288/mapk1-(mitogen-activated-protein-kinase-1)/ (accessed on 2 July 2024).
27. Available online: https://www.proteinatlas.org/ENSG00000100030-MAPK1 (accessed on 2 July 2024).
28. Li, S.; Ma, Y.M.; Zheng, P.S.; Zhang, P. GDF15 promotes the proliferation of cervical cancer cells by phosphorylating AKT1 and Erk1/2 through the receptor ErbB2. *J. Exp. Clin. Cancer Res.* **2018**, *37*, 80. [CrossRef] [PubMed]
29. Available online: https://www.uniprot.org/uniprotkb/P11473/entry#function (accessed on 2 July 2024).
30. Schulten, H.J.; Al-Mansouri, Z.; Baghallab, I.; Bagatian, N.; Subhi, O.; Karim, S.; Al-Aradati, H.; Al-Mutawa, A.; Johary, A.; Meccawy, A.A.; et al. Comparison of microarray expression profiles between follicular variant of papillary thyroid carcinomas and follicular adenomas of the thyroid. *BMC Genom.* **2015**, *16*, S7. [CrossRef] [PubMed]
31. Bi, C.; Li, B.; Du, L.; Wang, L.; Zhang, Y.; Cheng, Z.; Zhai, A. Vitamin D receptor, an important transcription factor associated with aldosterone-producing adenoma. *PLoS ONE* **2013**, *8*, e82309. [CrossRef] [PubMed]
32. DeSantis, K.A.; Robilotto, S.L.; Matson, M.; Kotb, N.M.; Lapierre, C.M.; Minhas, Z.; Leder, A.A.; Abdul, K.; Facteau, E.M.; Welsh, J. VDR in salivary gland homeostasis and cancer. *J. Steroid Biochem. Mol. Biol.* **2020**, *199*, 105600. [CrossRef]
33. Larriba, M.J.; Ordóñez-Morán, P.; Chicote, I.; Martín-Fernández, G.; Puig, I.; Muñoz, A.; Pálmer, H.G. Vitamin D receptor deficiency enhances Wnt/β-catenin signaling and tumor burden in colon cancer. *PLoS ONE* **2011**, *6*, e23524. [CrossRef]
34. Twigt, B.A.; Scholten, A.; Valk, G.D.; Rinkes, I.H.B.; Vriens, M.R. Differences between sporadic and MEN related primary hyperparathyroidism; clinical expression, preoperative workup, operative strategy and follow-up. *Orphanet. J. Rare Dis.* **2013**, *8*, 50. [CrossRef]
35. Gupta, G.K.; Gupta, G.K.; Agrawal, T.; Agrawal, T.; Pilichowska, M. Immunohistochemical expression of vitamin D receptor and forkhead box P3 in classic Hodgkin lymphoma: Correlation with clinical and pathologic findings. *BMC Cancer* **2020**, *20*, 535. [CrossRef] [PubMed]
36. Liu, H.; He, Y.; Beck, J.; da Silva Teixeira, S.; Harrison, K.; Xu, Y.; Sisley, S. Defining vitamin D receptor expression in the brain using a novel VDRCre mouse. *J. Comp. Neurol.* **2021**, *529*, 2362–2375. [CrossRef] [PubMed]
37. Rao, S.; Han, Z.H.; Phillips, E.R.; Palnitkar, S.; Parfitt, A.M. Reduced vitamin D receptor expression in parathyroid adenomas: Implications for pathogenesis. *Clin. Endocrinol.* **2000**, *53*, 373–381. [CrossRef]
38. Rehab Mohamed Sharaf, E.; Basma Mostafa Mahmoud, A.; Samia Ibrahim, E.N.; Wesam Ismail, M. Expression of Vitamin D Receptor (VDR) in Urinary Bladder Carcinoma: Immunohistochemical and Histopathological Study. *Int. J. Pathol. Clin. Res.* **2022**, *8*, 139. [CrossRef]
39. An, H.J.; Song, D.H. Displacement of Vitamin D receptor is related to lower histological grade of endometrioid carcinoma. *Anticancer. Res.* **2019**, *39*, 4143–4147. [CrossRef]

Case Report

Pitfalls of DualTracer 99m-Technetium (Tc) Pertechnetate and Sestamibi Scintigraphy before Parathyroidectomy: Between Primary-Hyperparathyroidism-Associated Parathyroid Tumour and Ectopic Thyroid Tissue

Mara Carsote [1,2], Mihaela Stanciu [3,*], Florina Ligia Popa [4,*], Oana-Claudia Sima [2,5], Eugenia Petrova [1,2], Anca-Pati Cucu [5,6] and Claudiu Nistor [6,7]

1   Department of Endocrinology, Carol Davila University of Medicine and Pharmacy, 050474 Bucharest, Romania; carsote_m@hotmail.com (M.C.); jekined@yahoo.com (E.P.)
2   Department of Clinical Endocrinology V, C.I. Parhon National Institute of Endocrinology, 020021 Bucharest, Romania; oanaclaudia1@yahoo.com
3   Department of Endocrinology, Faculty of Medicine, "Lucian Blaga" University of Sibiu, 550024 Sibiu, Romania
4   Department of Physical Medicine and Rehabilitation, Faculty of Medicine, "Lucian Blaga" University of Sibiu, 550024 Sibiu, Romania
5   PhD Doctoral School, Carol Davila University of Medicine and Pharmacy, 020021 Bucharest, Romania; ancutapati@gmail.com
6   Thoracic Surgery Department, Dr. Carol Davila Central Military Emergency University Hospital, 010242 Bucharest, Romania; ncd58@yahoo.com
7   Department 4—Cardio-Thoracic Pathology, Thoracic Surgery II Discipline, Carol Davila University of Medicine and Pharmacy, 050474 Bucharest, Romania
*   Correspondence: mihaela.stanciu@yahoo.com (M.S.); florina.popa@yahoo.com (F.L.P.)

Citation: Carsote, M.; Stanciu, M.; Popa, F.L.; Sima, O.-C.; Petrova, E.; Cucu, A.-P.; Nistor, C. Pitfalls of DualTracer 99m-Technetium (Tc) Pertechnetate and Sestamibi Scintigraphy before Parathyroidectomy: Between Primary-Hyperparathyroidism-Associated Parathyroid Tumour and Ectopic Thyroid Tissue. *Medicina* 2024, *60*, 15. https://doi.org/10.3390/medicina60010015

Academic Editors: Fumio Otsuka and Jin Wook Yi

Received: 15 October 2023
Revised: 27 November 2023
Accepted: 19 December 2023
Published: 21 December 2023

Abstract: Diagnosis of primary hyperparathyroidism (PHP) is based on blood assessments in terms of synchronous high calcium and PTH (parathormone), but further management, particularly parathyroid surgery that provides the disease cure in 95–99% of cases, requires an adequate localisation of the parathyroid tumour/tumours as the originating source, with ultrasound and 99m-Technetium (99m-Tc) sestamibi scintigraphy being the most widely used. We aimed to introduce an adult female case diagnosed with PHP displaying unexpected intra-operative findings (ectopic thyroid tissue) in relation to concordant pre-operatory imaging modalities (ultrasound + dual-phase 99m-Tc pertechnetate and sestamibi scintigraphy + computed tomography) that indicated bilateral inferior parathyroid tumours. A sudden drop in PTH following the removal of the first tumour was the clue for performing an extemporaneous exam for the second mass that turned out to be non-malignant ectopic thyroid tissue. We overviewed some major aspects starting from this case in point: the potential pitfalls of pre-operatory imaging in PHP; the concordance/discordance of pre-parathyroidectomy localisation modalities; the need of using an additional intra-operative procedure; and the clues of providing a distinction between pathological parathyroids and thyroid tissue. This was a case of adult PHP, whereas triple localisation methods were used before parathyroidectomy, showing concordant results; however, the second parathyroid adenoma was a false positive image and an ectopic thyroid tissue was confirmed. The pre-operatory index of suspicion was non-existent in this patient. Hybrid imaging modalities are most probably required if both thyroid and parathyroid anomalies are suspected, but, essentially, awareness of the potential pitfalls is mandatory from the endocrine and surgical perspectives. Current gaps in imaging knowledge to guide us in this area are expected to be solved by the significant progress in functional imaging modalities. However, the act of surgery, including the decision of a PTH assay or extemporaneous exam (as seen in our case), represents the key to a successful removal procedure. Moreover, many parathyroid surgeons may currently perform 4-gland exploration routinely, precisely to avoid the shortcomings of preoperative localisation.

Keywords: thyroid; parathyroid; 99m-Technetium; scintigraphy; thyroidectomy; parathyroidectomy; ectopic; primary hyperparathyroidism

1. Introduction

Diagnosis of primary hyperparathyroidism (PHP) is based on blood assessments in terms of synchronous high calcium and PTH (parathormone), but further management, particularly parathyroid surgery that provides the disease cure in 95–99% of cases, requires an adequate localisation of the parathyroid tumour/tumours as the originating source of the hormonal and biochemistry anomalies [1,2].

Parathyroid imaging represents a mandatory pre-operatory step and a large spectrum of tools are currently available, starting with the mostly frequently used, namely cervical ultrasound. While echography has the advantage of being non-expensive and not representing a source of irradiation, other techniques use different radiation doses such as parathyroid scintigraphy and four-dimensional (4D) computed tomography (CT) scans. Also, an enhancement in the baseline features as provided by the neck ultrasound may be provided by 4D MRI (magnetic resonance imaging) [3–5].

99m-Technetium (99m-Tc) is used as a tracer in relation to the parathyroid uptake (regardless if ectopic or orthotopic) and 99m-Tc sestamibi is commonly applicable, for instance, as single-photon scintigraphy with 99mTc-sestamibi or a scintigraphy with a dual tracer (99m-Tc pertechnetate and 99m-Tc sestamibi), with either variant being associated or not with thyroid subtraction [6,7].

99m-Tc scintigraphy represents a key imaging tool to address the parathyroid glands and, in most countries, almost every patient that is referred to parathyroidectomy had underwent at least one imaging procedure before surgery, mostly ultrasound in addition to 99m-Tc sestamibi scintigraphy [8,9]. Moreover, the results may be enhanced by adding SPECT (single-photon-emission computed tomography) or SPECT/CT regardless of an orthotopic or ectopic parathyroid tumour [10,11]. The underlying pathway of this molecular imaging is represented by the detection of the mitochondrial content at the level of parathyroid tissue, noting that hyper-functional parathyroid glands in PHP are associated with a high-oxyphil-cell-related abundant mitochondrial count [12,13]. The molecular imaging approach brings more functional information when compared to structural (standard) techniques such as CT or MRI. Further on, PET/CT (positron emission tomography), for instance, with ^{18}F-Fluorocholine or ^{11}C-methionine and PET/MRI (which requires lower radiation dose than PET/CT or CT scan), provides useful dynamic captures [7,14,15]. Additionally, angiographic selective venous sampling represents an alternative to the multimodal diagnosis approaches in the parathyroid field, but its exact placement among other modalities currently represents an emergent issue to be solved [16,17]. Importantly, this is an invasive tool that is largely reserved for re-operative cases in which imaging has not identified a target.

However, overall, despite recent progress in this specific area, there is no standard (unanimous) consensus with concern to the optimal imaging diagnosis before parathyroidectomy, with the applied techniques largely depending on one centre experience, the availability/costs of these procedures and also an individualised approach in relation to the clinical circumstances of a patient.

Of note, the clinical decision involving both endocrinology and surgery levels remains the core of this multidisciplinary management. While a skilled surgeon relates to the optimum parathyroidectomy outcome with regard to PHP and secondary (tertiary) hyperparathyroidism relates to the post-operatory success rate, complications and operation time, pre-operatory imaging diagnosis is mostly helpful, also allowing a minimally invasive/selective procedure, thus playing a role in the modern era in the parathyroid domain since this minimal surgical act in addition to the pre-operatory localisation of the tumour provides similar cure rates to bilateral neck exploration [18–20]. Nevertheless, many parathyroid surgeons may currently perform 4-gland exploration routinely, precisely to avoid the shortcomings of preoperative localisation.

Aim

We aimed to introduce the case of an adult female diagnosed with PHP displaying unexpected intra-operatory findings in relation to pre-operatory imaging modalities.

2. Method

This was a case report. The medical aspects were retrospectively registered. Biochemistry, hormonal and imaging panels were introduced, as well as intra-operatory (surgical) aspects. A brief literature overview is performed in the Discussion section with regard to potential pitfalls of pre-surgery imagery scans, particularly 99m-Tc scintigraphy in relation to pre-operatory concordant imaging results, and with its potential of differentiating between a thyroid and a parathyroid tissue.

3. Clinical Vignette

This was a non-smoking 56-year-old lady who was admitted after an accidental detection of increased serum calcium amid a routine evaluation that was conducted by her primary care physician. She had surgical menopause since the age of 40 for a benign uterine tumour and did not receive any hormone replacement therapy afterwards. Her family history was irrelevant; she has been known to have mild high blood pressure for the last decade (that was controlled under specific daily medication with beta blockers and calcium blocker—amlodipine). On admission, PHP was confirmed based on increased blood values of calcium and PTH (Table 1).

Table 1. Biochemistry and hormonal level on first admission for PHP on a 56-year-old female (* bone formation markers; ** bone resorption markers).

Parameter	Patient's Value	Normal Range
Total serum calcium (mg/dL)	11.3	8.5–10.2
Ionised calcium (mg/dL)	5.07	3.9–4.9
Total proteins(g/dL)	6.8	6.5–8.7
Phosphorus (mg/dL)	2.8	2.5–4.5
Creatinine (mg/dL)	0.77	0.5–1.1
Urea (mg/dL)	35	15–50
Hormones of mineral metabolism		
25-hydroxyvitamin D (ng/mL)	34	20–100
PTH (pg/mL)	103.1	15–65
Bone turnover markers		
Osteocalcin (ng/mL) *	48.2	15–46
Alkaline phosphatase(U/L) *	117	38–105
P1NP (ng/mL) *	103.1	14.28–58.92
CrossLaps (ng/mL) **	1.05	0.33–0.782

The patient had normal thyroid function and negative serum antibodies against thyroid. DXA (Dual-Energy X-ray Absorptiometry thorough a GE Lunar Prodigy device) confirmed osteoporosis (no prevalent vertebral fracture was detected at profile X-ray at the level of thoracic–lumbar spine) (Table 2).

Table 2. Central DXA (GE Lunar Prodigy) report in a menopausal female with PHP with osteoporosis confirmation based on T-score.

Region	Bone Mineral Density (g/cm^2)	T-Score (SD)	Z-Score (SD)
Lumbar spine L1–4	0.764	−3.5	−2.5
Femoral neck	0.640	−2.8	−1.9
Total hip	0.606	−3.3	−2.6
1/3 distal radius	0.592	−3.2	−2.7

A value of lumbar DXA-based TBS (trabecular bone score) of 1349 (iNsight) was considered a partially degraded microarchitecture (yet, close to the normal cut off of at least 1350) [21]. No kidney stone was detected at abdominal ultrasound, nor any component of multiple endocrine neoplasia syndromes. Noting the PHP–osteoporosis (of course, with an additional early-menopause-related component) and a value of serum total calcium above +1 mg/dL above the upper normal limit, the patient had an indication of parathyroidectomy [22]; thus, localisation scans were performed starting with neck ultrasound that suggested a right inferior parathyroid tumour (Figure 1).

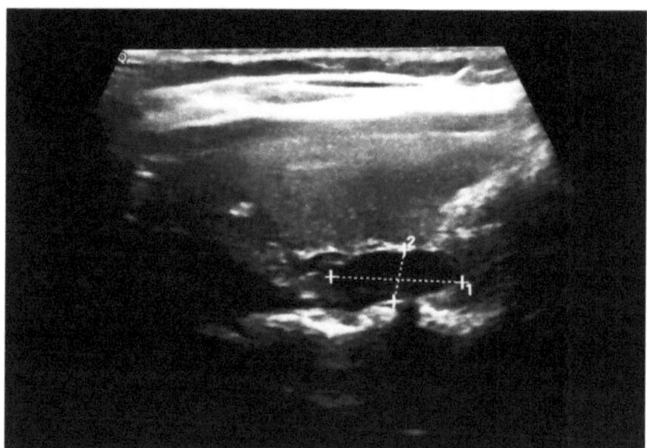

Figure 1. Neck ultrasound showing a right thyroid lobe of 2/2/5 cm; a left thyroid lobe of 2/1/4 cm with hypoechoic, inhomogeneous pattern; and, posterior and inferior to the right thyroid lobe, a hypoechoic nodule, inhomogeneous, with no vascularisation, of 1.56/0.6/0.7 cm, suggestive for a parathyroid adenoma.

99m-Tc parathyroid scintigraphy was performed (a dual technique with 99mTcpertechnetate and 99m-Tcsestamibi) and showed two areas of increased late uptake of 99m-Tc sestamibi at the level of the inferior pole of the left thyroid lobe and of the right thyroid lobe, respectively, suggesting two bilateral inferior parathyroid tumours (Figure 2).

These findings were confirmed at intravenous contrast CT scan (Figure 3).

Due to the imaging diagnosis of multi-glandular disease, other hormonal assays were carried out (such as calcitonin or metanephrines/normetanephrines) but were found negative for a diagnosis of multiple endocrine neoplasia. After performing parathyroid scintigraphy and a CT scan, neck ultrasound was re-performed and a potential mass of around 1 cm at the level of the left latero-cervical area was suspected as being a parathyroid tumour.

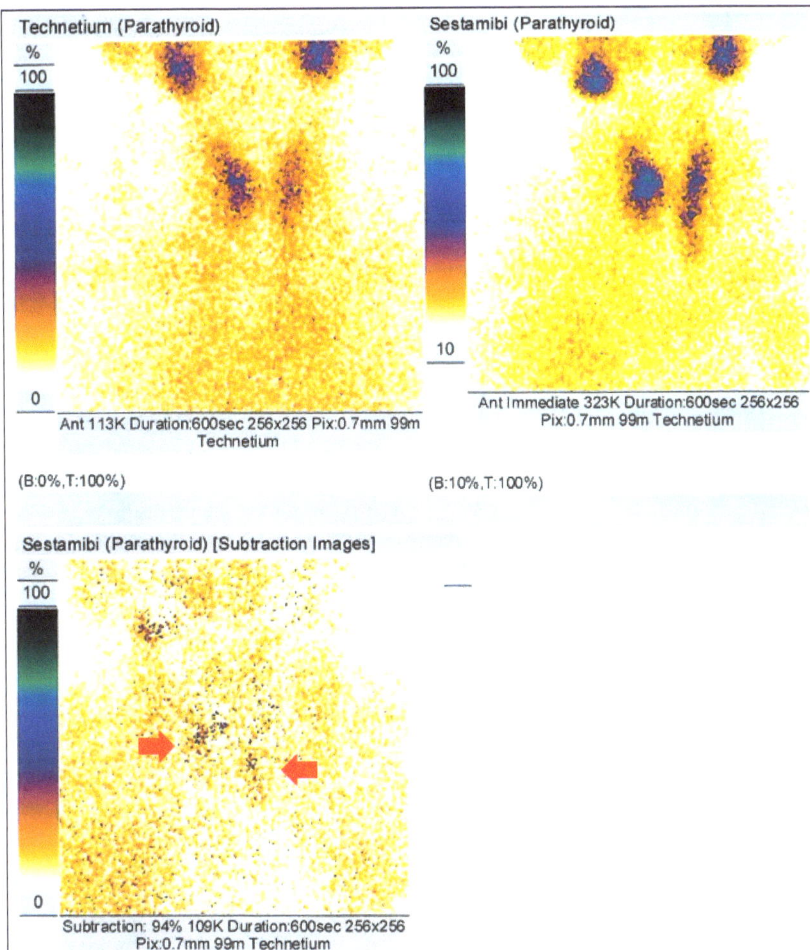

Figure 2. 99m-Tc parathyroid scintigraphy with 99m-Tc pertechnetate (185 MBq) and 99m-Tc sestamibi (740 mBq; effective dose of 9.06 mSv); 94% subtraction captures (at 60 min show two late-uptake areas at the level of left and right inferior thyroid lobes, suggestive for parathyroid adenomas (red arrow).

Herein, the patient was offered 5 mg of intravenous zoledronate in order to control hypercalcaemia (with a post-injection decrease in total calcium to 10.4 mg/dL) and to serve as an anti-osteoporotic regime (5 mg zoledronate per year) in addition to daily cholecalciferol of 1000 UI.

The patient was further referred to a one-time parathyroidectomy of both tumours. Firstly, the right inferior parathyroid tumour was removed, followed by an intra-operatory PTH assay that showed a very low value. Since the high-PTH-originating source seemed to be this adenoma, an extemporaneous exam was performed for the second mass (that was initially suspected to be a synchronous parathyroid adenoma, as well) and revealed an ectopic thyroid tissue. Thus, an intra-operatory decision was made for further removal with post-operatory histological confirmation of this ectopic thyroid tissue (with dilated follicles, and colloid content, without atypia or malignant elements). The right para-tracheal tumour was confirmed post operation as being a parathyroid adenoma without vascular invasion (Figure 4).

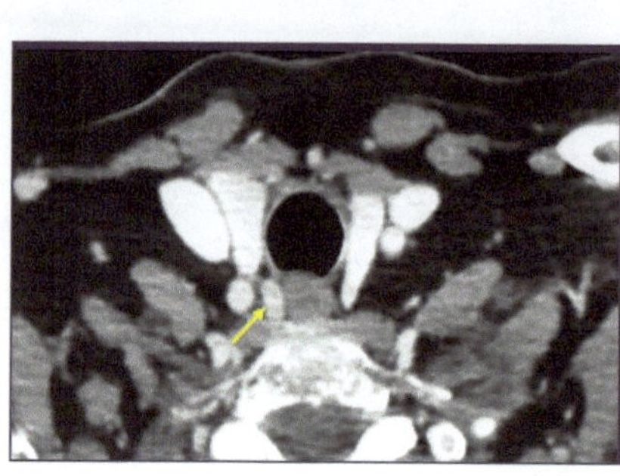 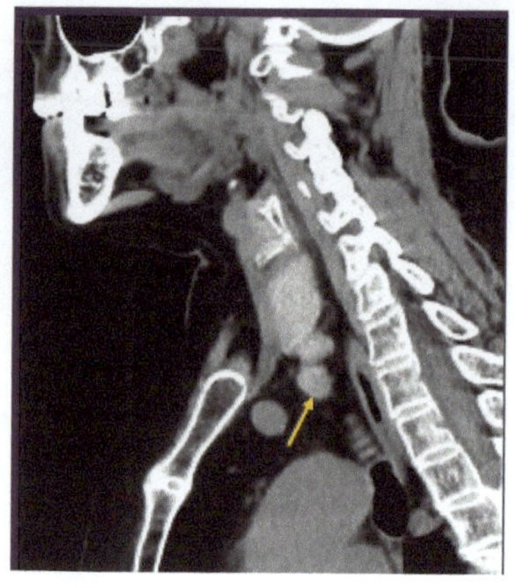

Figure 3. CT scan showing an oval-to-round, well-circumscribed, iodophile, slightly heterogeneous nodule at the right para-tracheal level (in the superior mediastinum–manubrium level) of 1.12/1.33/1.40 cm (yellow arrow; transversal plan), respectively, and an oval, well-circumscribed, iodophile nodule at latero-cervical, left para-oesophageal, clavicular level, of 0.56/1.27/1.72 cm (orange arrow; sagittal plan).

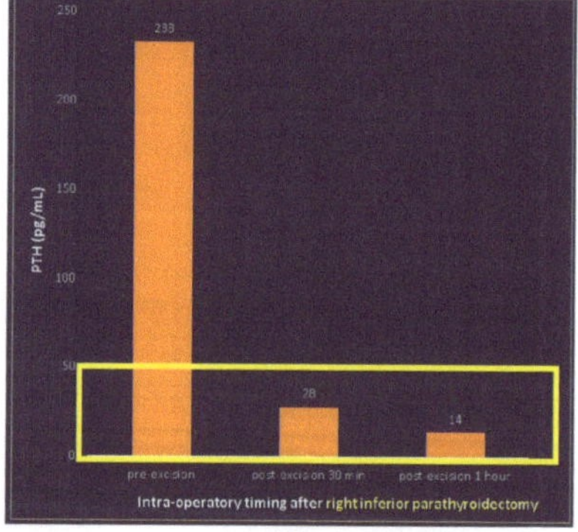

(A)

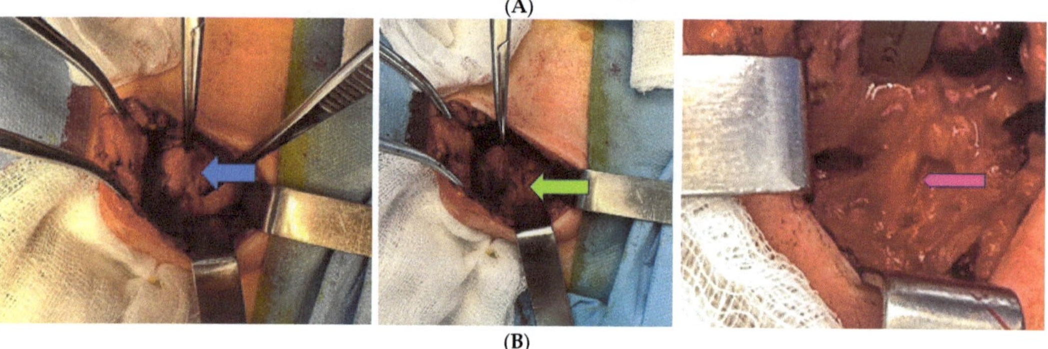

(B)

Figure 4. *Cont.*

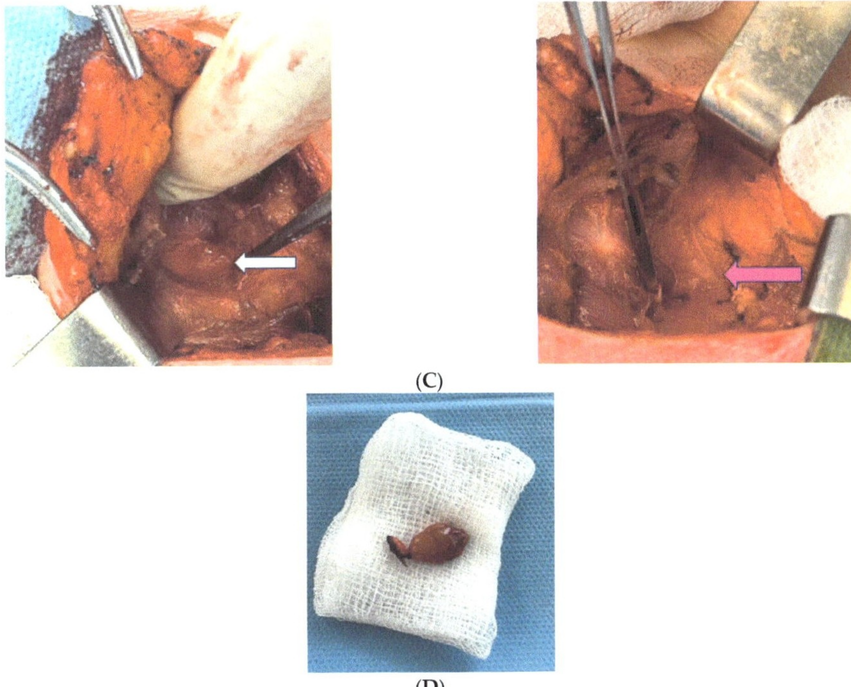

Figure 4. Surgical aspects: (**A**) PTH assays after right inferior parathyroid tumour removal. (**B,C**) Intra-operatory captures: (**B**) right inferior parathyroid tumour (blue arrow); right inferior thyroid pedicle (green arrow); right recurrent laryngeal nerve (pink arrow). (**C**) Ectopic (upper mediastinal) thyroid tissue that, pre-operation, mimicked an additional left parathyroid tumour (white arrow); left recurrent laryngeal nerve (pink arrow). (**D**) Post-operatory specimen: macroscopic aspect of the right inferior parathyroid adenoma.

The post-surgery outcome was uneventful; the patient experienced a normalisation of calcium and PTH levels without any influence on thyroid function. The post-parathyroidectomy 1-month PTH was 64 pg/mL (Figure 5). The lady continued vitamin D replacement and a periodic check-up is required including annual DXA.

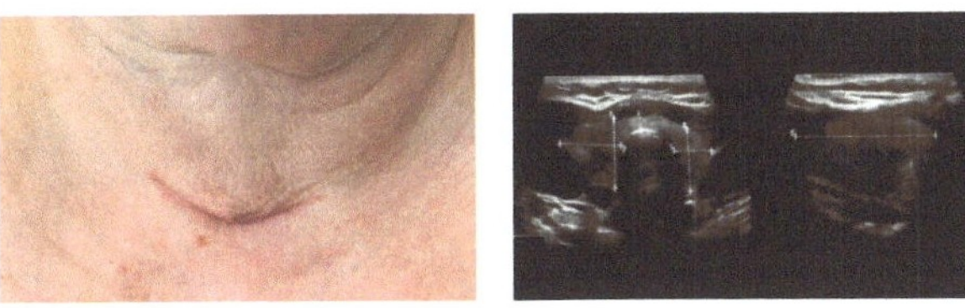

Figure 5. Post-parathyroidectomy scar within the first few weeks (**left**); anterior neck ultrasound showing thyroid features similar to pre-operatory findings and no remnants at the level, whereas both masses have been removed (**right**).

4. Discussion

We highlight some major aspects starting from this case in point: the potential pitfalls of pre-operatory imaging in PHP; the concordance/discordance of pre-parathyroidectomy localisation modalities; and the need of using an additional intra-operatory procedure in certain cases, which are clues for providing a distinction between pathological parathyroids and thyroid tissue.

4.1. Imaging Pitfalls in PHP

This case reveals an interesting and challenging aspect: incongruent findings between pre-operatory imaging captures pinpointing a double parathyroid adenoma in terms of 99m-Tc pertechnetate (TcO4)–sestamibi plan scintigraphy and a CT exam, on the one hand, and the intra/post-operatory confirmation with regard to one of these masses, one the other hand (one of the presumably "parathyroid" tumours proved to be an ectopic thyroid tissue with no malignancy features). An intra-operatory decision was mandatory amid lowering PTH values after the resection of the first mass. Since Tc scintigraphy was found positive on both sides, a potential non-functional parathyroid tumour was suspected on the left (thus explaining the PTH normalisation), while an ectopic thyroid tissue seemed more likely based on its macroscopic features (and this was finally confirmed). The removal of this second mass was based on the fact that it was not a normal parathyroid, nor a thyroid gland; hence, it represented an individual decision in this lady's case, not a matter of guidance. However, in addition to the results upon the first-hand histological (extemporaneous) report, a consultation with her current endocrinologist was performed amidst surgery in order to proceed with the second tumour removal.

Generally, a good localisation study before performing parathyroidectomy is helpful for the surgeon and avoids unnecessary prolonged surgery time or even a redo of parathyroidectomy, thus allowing a minimally invasive approach (and a shorter hospitalisation stay and increased patient' comfort) [12,13]. Also, an adequate pre-operatory localisation, regardless of the methods, improves the PTH values within the first minutes after tumour removal (as seen in this mentioned vignette) [23].

Generally, the use of the 99m-Tc sestamibi-based scan is traditionally a part of the pre-parathyroidectomy preparation panel (740–924 MBq). A 99m-Tc radiotracer may be administered via injection (intravenously) or orally and it may also be used for thyroid scintigraphy (sodium pertechnetate), both for adults (1–10 mCi) and children (60–80 µCi/kg). Increased homogeneous tracer uptake is suggestive for thyroid/parathyroid tissue in dual scintigraphy (sestamibi and pertechnetate), mostly depending on the uptake timing (early for thyroid and late/delayed for parathyroid at the moment when thyroid images are already washed out) [24].

However, despite performing imaging techniques, some pitfalls cannot be predicted, especially if concordant localisation results are found with different methods (for instance, in our case, CT combined with 99m-Tc plan scintigraphy); thus, the appeal of using an additional technique did not seem justified [12,13].

A spectrum of bias relates to the clinical, hormonal and anatomical aspects in PHP as well as technical issues, including their availability/access in one centre (and associated experience of the medical team). Inter-observer differences have been reported with concern to Tc-99m-based scintigraphy, but, generally, ultrasound is regarded as the most commonly recognised subject of pitfalls [25]. A higher rate of localisation failure was also described in ectopic intra-thyroid parathyroid adenomas [26]. Also, a lower PTH value at baseline is correlated with a negative finding at 99m-Tc scintigraphy; thus, in these cases a combination of several modalities is required from the start [27]. A multi-glandular disease (generally representing 20% of the adult cases with PHP, particularly those with a strong genetic background) associates with a higher rate of identification failure regardless of the pre-surgery functional imaging modalities (particularly when ectopic pathological glands are also involved) [28]. 99m-Tc sestamibi plan scintigraphy has been used not only in PHP, but also in secondary (chronic kidney disease-associated) hyperparathyroidism, with the sensitivity for paediatric cases being lower than seen in the adult population (for example, 40% versus 70%) with an increased rate of reduced radiotracer uptake at the thyroid level (42% versus 2%) [29]. The pre-surgery detection rate is generally higher in PHP than in the secondary type [30].

Another dual-phase scintigraphy, namely dual tracer (99m-Tc and 123 Iodine), provides simultaneous images that avoid the subtraction-related artefacts and might prove beneficial in multi-glandular parathyroid disease in addition to thyroid gland-related information [31,32]. The traditional issues of blocked thyroid uptake after recent iodine exposure (as seen after using iodine contrast CT) or in individuals under chronic levothyroxine replacement therapy should be taken into consideration when using iodine scintigraphy [33]. Of course, there is a standard issue of availability; for instance, at the moment when we evaluated the patient, iodine-based scintigraphy was not available.

In order to enhance the performance of 99m-Tc dual-phase plan scintigraphy, an alternative is represented by 99m-Tc-MIBI SPECT/CT fusion imaging [34]. SPECT/CT offers an increased sensitivity and accuracy for location diagnosis and it became a first-line option in some centres [35] (but not in ours) or it may be applied as an elegant alternative in difficult cases such as suspected recurrent tumours or carcinomas [36]. One limit was found to be a reduced pathological tumour weight (false negative results) [37] and some data showed a decreased rate of localisation in normocalcaemic PHP versus PHP, associating a classical presentation with high levels of serum calcium that might involve a smaller parathyroid gland. On the contrary, a hypercalcaemia-related inhibitory effect of the radiotracer uptake has been reported, too [12,13,38]. Also, the use of calcium blockers (as in our case according to patient's medical records in order to control the arterial hypertension) and calcimimetics such as cinacalcet may reduce the 99m-Tc MIBI uptake [12,13] (Figure 6).

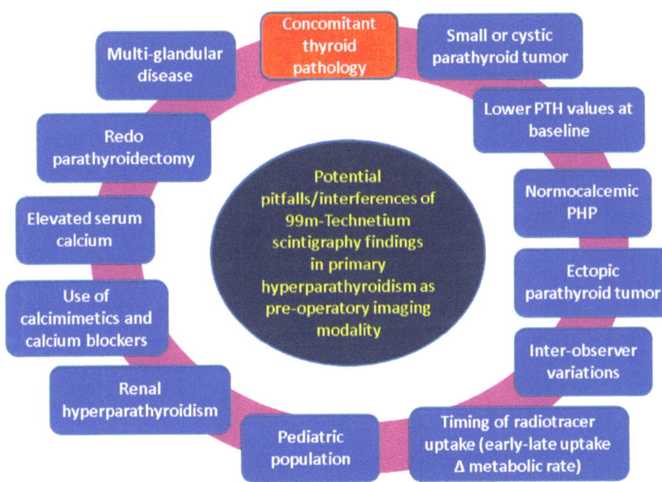

Figure 6. Sneak peak of potential pitfalls when addressing the results of 99m-Tc sestamibi scintigraphy in PHP [1–80].

4.2. Concordance of Pre-Operatory Localisation Studies in PHP

This was a case of adult PHP, whereas triple localisation methods were used before parathyroidectomy, showing concordant results; however, the second parathyroid adenoma was a false positive image. In addition to 99m-Tc-based assessment, a standard CT scan was conducted in this mentioned situation. Despite recent progress in the era of imaging in PHP, CT remains a mostly used and feasible approach as part of a routine exam in real-life medicine [39,40]. Dual-phase (non-enhanced and arterial) CT protocols increase the accuracy of results, in both the adult and paediatric (but less accurate) population with PHP [41]. Of note, CT should be avoided if possible in children, especially as non-radiating techniques are available. Alternatively, dual-energy CT (DECT) showed similar accuracy with conventional imaging techniques [42]. Notably, the concordance of pre-operatory localisation diagnosis did not raise the suspicion of further using an additional imaging technique in this instance.

Some data suggested that PTH levels that are not extremely elevated might mislead the interpretation of 99m-Tc-based scintigraphy. For example, one study correlated the cut off values of serum calcium and PTH with positive Tc-99m-MIBI (methoxyisobutylisonitrile) parathyroid scintigraphy (and positive parathyroid subtraction) in patients diagnosed with PHP (median age of 60, between 22 and 78 years). PTH (not serum calcium) statistically significantly correlated with 99m-Tc scintigraphy findings [43]. This implies that cases with a PTH level around 100 pg/mL (as seen in our case) should be taken into consideration in interpreting scintigraphy.

Recently, 11C-methionine, 11C-choline or 18F-fluorocholine PET/CT showed encouraging outcomes, especially in patients who underwent unsuccessful parathyroidectomy or had negative or discordant imaging results pre-surgery, with these patients being alternatively candidates to SPECT/CT, as well [44–47]. Head-to-head studies (versus 99m-Tc sestamibi SPECT/CT) showed comparable results despite using a different tracer for parathyroid glands [48] or even an improvement in detection rate, for instance, from 88% for 99m-Tc-sestamibi SPECT/CT to 98% according to one recent study [49] or a sensitivity of 99% for 18F-fluorocholine PET/CT versus 75% for ultrasound versus 65% for 99m-Tc sestamibi scintigraphy versus 89.9% for ultrasound combined with this type of scintigraphy according to another cohort from 2022 [50].

As mentioned, a lower detection rate was described in multi-glandular disease/hyperplasia versus single-gland involvement [48,49]. Alternatively, 18F-fluorocholine PET/ultrasound fusion imaging might bring supplementary benefits [51]. 11C-choline PET/CT might help in cases with negative/discordant data when using traditional methods such as CT and/or Tc scintigraphy, with approximately 93% of such cases being true positive [52]. 18F-fluorocholine PET/MRI has recently been proven to have a similar or higher accuracy than 99m-Tc sestamibi scintigraphy, with the association of these two modalities being currently recommended [53]. Currently, there is no such thing as a unique algorithm of pre-operatory functional multimodal imaging and the use of PET/CT or PET/MRI in PHP is not standardised yet, largely depending on their availably in one centre and local protocols [54].

An alternative to the mentioned pre-operatory imaging methods was proposed to be ultrasound-guided parathyroid fine-needle aspiration (with PTH washout), but so far, there are not so many large studies to address this issue, which is not conventionally approved in many centres [55]. However, one study from 2023 showed a similar positive predictive value with 99m-Tc sestamibi scintigraphy; thus, it might become a first-line imaging modality in a selective subgroup of subjects with PHP [56]. Also, the rate of non-localisation in patients who consequently required a redo of parathyroidectomy is higher; thus, parathyroid vein sampling was proposed as an intra-operatory additional method for those subjects with a negative pre-operatory imaging diagnosis in recurrent/persistent PHP [57]. Of note, intra-operatory PTH monitoring or assessments (as we used in our case, too) represent an essential clue for the parathyroidectomy success [58,59].

Another approach involves an alternative to radiologist-based—or endocrinologist-based—pre-operatory neck ultrasound, namely a similar pre-incision ultrasound that is carried out by the surgeon on the operating table while the patient is under general anaesthesia just before the actual parathyroid removal. This intra-operatory "before skin incision" localisation procedure might enhance the results of pre-operatory echography, thus allowing a minimally invasive parathyroidectomy. However, the implementation of such a protocol largely depends on each centre standards rather than being a guideline recommendation [60–63]. Overall, negative or discordant pre-operatory imaging results in PHP do not exclude a successful parathyroidectomy in the hands of a skilled surgeon [44,45].

4.3. Ectopic Thyroid Tissue

While false negative results at 99m-Tc sestamibi scintigraphy may involve small parathyroid tumours or parathyroid cysts, false positive results relate to abnormal tracer uptake at the thyroid level (in our case, only at ectopic tissue, not at normal thyroid gland). The fact that 99m-Tc MIBI is up-taken in highly metabolic lesions might imply a different cellular turnover in ectopic, not orthotopic, thyroids, as has similarly been reported (false positive results) in different types of head and neck cancers, even breast and lung neoplasia [64–67]. Hence, in this case, the incidental detection of the ectopic thyroid was globally due to the first diagnosis of PHP and associated management, while its actual recognition started during surgery from an unexpected PTH drop after the excision of the first tumour. We identified a similarity with another case published in 2019: this was a 52-year-old male confirmed with PHP. Pre-operatory MIBI SPECT/CT confirmed a late uptake on two focal areas at the level of the left thyroid lobe, but the post-surgery histological report showed that only one was a parathyroid adenoma and the other was a thyroid hyperplasia. Notably, pre-operatory 99m-Tc MIBI planar scintigraphy was not relevant, only showing positive at SPECT/CT; thus, we may conclude that false positive images due to synchronous concurrent thyroid conditions may be detected even with regard to advanced imaging modalities [68].

On the other hand, with concern to parathyroid tumours, pre-operatory false negative results involve between 5.7 and 25% of the patients who perform 99m-Tc sestamibi SPECT scintigraphy, and they are related to the limits of resolution of the technique or a prolonged time from scintigraphy to surgery [69,70]. Also, in patients with large goitres, there is a higher rate of false negative results at ultrasound and 99m-Tc sestamibi scintigraphy [71]. As mentioned, a negative parathyroid localisation result (even caused by thyroid issues) might be overcome by an experienced surgeon in the field of parathyroid tumours, so-called "the sestamibi paradox" [72].

A similar case regarding both parathyroid and thyroid pathological findings was reported in 2023 on a subject who underwent 99mTc-sestamibi scintigraphy: the lack of orthotopic uptake at the level of the thyroid area was associated with an ectopic thyroid tissue that was identified at the lingual level, while, synchronously, there was an ectopic mediastinal parathyroid tumour [73]. Another very rare scenario involves the concomitant diagnosis of PHP and giant-goitre-associated thyrotoxicosis; in this situation, hypercalcaemia-related hyperthyroidism may mask the biological recognition of PHP and 99m-Tc sestamibi scintigraphy might be found as false negative at baseline [74]. Also, 99m-Tc scintigraphy has been used for thyroid hemi-agenesis, showing an increased unilateral uptake amid this interesting developmental disease of the gland [75,76].

Finally, some potential implications may be related to using a dual tracer at 99m-Tc scintigraphy. The 99m-Tc pertechnetate radiotracer may be up-taken by the thyroid and other organs such as the gastric mucosa [77]. Non-iodine-based methods for the functional and anatomic study of the thyroid gland include not only 99m-Tc sestamibi scintigraphy, but also 18F-FDG (^{18}F-fluoro-2-deoxy-d-glucose) PET/CT [78]. Further on, 99m-Tc sestamibi has been applied (via different quantitative parameters) in order to stratify the malignancy risk of the cold nodules (that do not uptake 99m-Tc TcO4) with indeterminate results following a thyroid-fine-needle-aspiration-associated cytological exam [79–81]. However, in our case, the late uptake amid dual-tracer plan scintigraphy was registered only at the level of ectopic thyroid tissue synchronously with the parathyroid adenoma, and not at the physiological thyroid. Hybrid imaging modalities are most probably required in both thyroid and parathyroid anomalies, but, essentially, awareness of the potential pitfalls is mandatory from the clinical and surgical perspectives [82].

5. Conclusions

This was a case of adult PHP, whereas triple localisation methods were used before parathyroidectomy, showing concordant results. However, the second parathyroid adenoma was a false positive image and an ectopic thyroid tissue was confirmed. The pre-operatory index of suspicion was non-existent in this patient. Thus, we may conclude that in a selected subgroup of individuals, hybrid imaging modalities might prove useful if both thyroid and parathyroid conditions are present, but, essentially, awareness of such potential pitfalls is mandatory from the endocrine and surgical perspectives. Current gaps in imaging knowledge to guide us in these specific areas are expected to be solved by the significant progress in functional imaging modalities. However, the act of surgery, including the decision of a PTH assay or extemporaneous exam (as seen in our case), represents the major key to a successful removal procedure.

Author Contributions: Conceptualization, M.C., M.S., F.L.P., O.-C.S., E.P., A.-P.C. and C.N.; methodology, M.C., M.S., F.L.P., E.P., A.-P.C. and C.N.; software, M.C., M.S., F.L.P., E.P., A.-P.C. and C.N.; validation, M.C., M.S., F.L.P., O.-C.S., A.-P.C. and C.N.; formal analysis, M.C., M.S., F.L.P. and C.N.; investigation, M.C., M.S., O.-C.S., E.P. and C.N.; resources, M.C., M.S., F.L.P., A.-P.C. and C.N.; data curation, M.C., M.S., F.L.P., O.-C.S., E.P., A.-P.C. and C.N.; writing—original draft preparation, M.C., O.-C.S., E.P., A.-P.C. and C.N.; writing—review and editing, M.C., M.S., F.L.P. and C.N.; visualization, M.C., M.S., F.L.P. and C.N.; supervision, M.C., M.S., F.L.P. and C.N.; project administration, M.C., M.S. and C.N.; funding acquisition, M.S. All authors have read and agreed to the published version of the manuscript.

Funding: Project financed by Lucian Blaga University of Sibiu (Knowledge Transfer Center) & Hasso Plattner Foundation research grants LBUS-HPI-ERG-2023-05.

Institutional Review Board Statement: This study was approved by the Institutional Review Board of Dr. Carol Davila Central Military Emergency University Hospital, Bucharest, Romania (approval number: 608/28 June 2023).

Informed Consent Statement: Written informed consent has been obtained from the patient to publish this paper.

Data Availability Statement: The original data generated and analysed for this case presentation are included in the published article.

Acknowledgments: We thank the patient and all the medical and surgical teams that were involved in this case.

Conflicts of Interest: The authors declare no conflict of interest.

Abbreviations

CT	computed tomography
4D	four-dimensional
DXA	Dual-Energy X-ray Absorptiometry
DECT	dual-energy computed tomography
MIBI	methoxyisobutylisonitrile
18F-FDG	^{18}F-fluoro-2-deoxy-d-glucose
MRI	magnetic resonance imaging
PHP	primary hyperparathyroidism
PET	positron emission tomography
PTH	parathormone
SPECT	single-photon-emission computed tomography
Tc	Technetium
TBS	trabecular bone score

References

1. Zielke, A.; Smaxwil, C.A. Current approach in cases of persistence and recurrence of primary hyperparathyroidism. *Chirurgie* **2023**, *94*, 595–601. [CrossRef] [PubMed]
2. Bijnens, J.; Van den Bruel, A.; Vander Poorten, V.; Goethals, I.; Van Schandevyl, S.; Dick, C.; De Geeter, F. Retrospective real-life study on preoperative imaging for minimally invasive parathyroidectomy in primary hyperparathyroidism. *Sci. Rep.* **2022**, *12*, 17427. [CrossRef] [PubMed]
3. Morris, M.A.; Saboury, B.; Ahlman, M.; Malayeri, A.A.; Jones, E.C.; Chen, C.C.; Millo, C. Parathyroid Imaging: Past, Present, and Future. *Front. Endocrinol.* **2022**, *12*, 760419. [CrossRef] [PubMed]
4. Strauss, S.B.; Roytman, M.; Phillips, C.D. Parathyroid Imaging: Four-dimensional Computed Tomography, Sestamibi, and Ultrasonography. *Neuroimaging Clin. N. Am.* **2021**, *31*, 379–395. [CrossRef] [PubMed]
5. Cruz-Centeno, N.; Longoria-Dubocq, T.; Mendez-Latalladi, W. Efficacy of 4D CT Scan in Re-operative Parathyroid Surgery. *Am. Surg.* **2022**, *88*, 1549–1550. [CrossRef]
6. Lalonde, M.N.; Correia, R.D.; Syktiotis, G.P.; Schaefer, N.; Matter, M.; Prior, J.O. Parathyroid Imaging. *Semin. Nucl. Med.* **2023**, *53*, 490–502. [CrossRef] [PubMed]

7. Fiz, F.; Bottoni, G.; Massollo, M.; Trimboli, P.; Catrambone, U.; Bacigalupo, L.; Righi, S.; Treglia, G.; Imperiale, A.; Piccardo, A. [18F]F-Choline PET/CT and 4D-CT in the evaluation of primary hyperparathyroidism: Rivals or allies? *Q. J. Nucl. Med. Mol. Imaging* **2023**, *67*, 130–137. [CrossRef]
8. Fendrich, V.; Zahn, A. Localization diagnostics and operative strategy for the first intervention in sporadic primary hyperparathyroidism. *Chirurgie* **2023**, *94*, 573–579. [CrossRef]
9. Park, H.S.; Hong, N.; Jeong, J.J.; Yun, M.; Rhee, Y. Update on Preoperative Parathyroid Localization in Primary Hyperparathyroidism. *Endocrinol. Metab.* **2022**, *37*, 744–755. [CrossRef]
10. Abraham, B.M., Jr.; Sharkey, E.; Kwatampora, L.; Ranzinger, M.; von Holzen, U. Mediastinal Intrathymic Parathyroid Adenoma: A Case Report and Review of the Literature. *Cureus* **2023**, *15*, e42306. [CrossRef]
11. Özçevik, H.; Öner Tamam, M.; Tatoğlu, M.T.; Mülazımoğlu, M. Comparison of Planar Imaging Using Dual-phase Tc-99m-sestamibi Scintigraphy and Single Photon Emission Computed Tomography/Computed Tomography in Hyperparathyroidism. *Mol. Imaging Radionucl. Ther.* **2022**, *31*, 191–199. [CrossRef] [PubMed]
12. Petranović Ovčariček, P.; Giovanella, L.; Hindie, E.; Huellner, M.W.; Talbot, J.N.; Verburg, F.A. An essential practice summary of the new EANM guidelines for parathyroid imaging. *Q. J. Nucl. Med. Mol. Imaging* **2022**, *66*, 93–103. [CrossRef] [PubMed]
13. Petranović Ovčariček, P.; Giovanella, L.; Carrió Gasset, I.; Hindié, E.; Huellner, M.W.; Luster, M.; Piccardo, A.; Weber, T.; Talbot, J.N.; Verburg, F.A. The EANM practice guidelines for parathyroid imaging. *Eur. J. Nucl. Med. Mol. Imaging* **2021**, *48*, 2801–2822. [CrossRef] [PubMed]
14. Zarei, A.; Karthik, S.; Chowdhury, F.U.; Patel, C.N.; Scarsbrook, A.F.; Vaidyanathan, S. Multimodality imaging in primary hyperparathyroidism. *Clin. Radiol.* **2022**, *77*, e401–e416. [CrossRef] [PubMed]
15. Hanba, C.; Khariwala, S.S. What Is the Optimal Imaging Modality for Parathyroid Adenoma? *Laryngoscope* **2022**, *132*, 1508–1509. [CrossRef]
16. Gök, İ.; Şahbaz, N.A.; Akarsu, C.; Cem Dural, A.; Mert, M.; Erbahçeci Salık, F.A.; Çil, B.E.; Güzey, D.; Alış, H. The role of selective venous sampling in patients with non-localized primary hyperparathyroidism. *Turk. J. Surg.* **2020**, *36*, 164–171. [CrossRef]
17. Uludag, M. Preoperative Localization Studies in Primary Hyperparathyroidism. *Med. Bull. Sisli Etfal Hosp.* **2017**, *53*, 7–15. [CrossRef]
18. Rawat, A.; Grover, M.; Kataria, T.; Samdhani, S.; Mathur, S.; Sharma, B. Minimally Invasive Parathyroidectomy as the Surgical Management of Single Parathyroid Adenomas: A Tertiary Care Experience. *Indian J. Otolaryngol. Head Neck Surg.* **2023**, *75*, 271–277. [CrossRef]
19. Iwen, K.A.; Kußmann, J.; Fendrich, V.; Lindner, K.; Zahn, A. Accuracy of Parathyroid Adenoma Localization by Preoperative Ultrasound and Sestamibi in 1089 Patients with Primary Hyperparathyroidism. *World J Surg.* **2022**, *46*, 2197–2205. [CrossRef]
20. Nistor, C.E.; Stanciu-Găvan, C.; Vasilescu, F.; Dumitru, A.V.; Ciuche, A. Attitude of the surgical approach in hyperparathyroidism: A retrospective study. *Exp. Ther. Med.* **2021**, *22*, 959. [CrossRef]
21. Silva, B.C.; Broy, S.B.; Boutroy, S.; Schousboe, J.T.; Shepherd, J.A.; Leslie, W.D. Fracture Risk Prediction by Non-BMD DXA Measures: The 2015 ISCD Official Positions Part 2: Trabecular Bone Score. *J. Clin. Densitom.* **2015**, *18*, 309–330. [CrossRef] [PubMed]
22. Iacobone, M.; Scerrino, G.; Palazzo, F.F. Parathyroid surgery: An evidence-based volume—Outcomes analysis: European Society of Endocrine Surgeons (ESES) positional statement. *Langenbecks Arch. Surg.* **2019**, *404*, 919–927. [CrossRef] [PubMed]
23. Bilen, N.; Gokalp, M.A.; Yilmaz, L.; Aytekin, A.; Baskonus, I. Analysis of intraoperative laboratory measurements and imaging techniques such as Tc-99 m-MIBI SPECT/CT, 18F-fluorocholine PET/CT and ultrasound in patients operated with prediagnosis of parathyroid adenoma. *Ir. J. Med. Sci.* **2023**, *192*, 1695–1702. [CrossRef] [PubMed]
24. Maccora, D.; Fortini, D.; Moroni, R.; Sprecacenere, G.; Annunziata, S.; Bruno, I. Comparison between MIBI-based radiopharmaceuticals for parathyroid scintigraphy: Quantitative evaluation and correlation with clinical-laboratory parameters. *J. Endocrinol. Investig.* **2022**, *45*, 2139–2147. [CrossRef] [PubMed]
25. Baumgarten, J.; Happel, C.; Ackermann, H.; Grünwald, F. Evaluation of intra- and interobserver agreement of Technetium-99m-sestamibi imaging in cold thyroid nodules. *Nuklearmedizin* **2017**, *56*, 132–138. [CrossRef] [PubMed]
26. Gowrishankar, S.V.; Bidaye, R.; Das, T.; Majcher, V.; Fish, B.; Casey, R.; Masterson, L. Intrathyroidal parathyroid adenomas: Scoping review on clinical presentation, preoperative localization, and surgical treatment. *Head Neck* **2023**, *45*, 706–720. [CrossRef]
27. Neuberger, M.; Dropmann, J.A.; Kleespies, A.; Fuerst, H. Determinants and clinical significance of negative scintigraphic findings in primary hyperparathyroidism: A retrospective observational study. *Nuklearmedizin* **2022**, *61*, 440–448. [CrossRef]
28. Kowalski, G.; Buła, G.; Bednarczyk, A.; Gawrychowska, A.; Gawrychowski, J. Multiglandular Parathyroid Disease. *Life* **2022**, *12*, 1286. [CrossRef]
29. Aktaş, A.; Baskin, E.; Gençoğlu, E.A.; Çolak, T. Comparison of parathyroid scintigraphy findings in pediatric and adult patients with secondary hyperparathyroidism. *Nucl. Med. Commun.* **2023**, *44*, 860–863. [CrossRef]
30. Wang, Y.; Liu, Y.; Li, N.; Zhang, W. Comparison of biochemical markers and technetium ^{99m}methoxyisobutylisonitrile imaging in primary and secondary hyperparathyroidism. *Front. Endocrinol.* **2023**, *14*, 1094689. [CrossRef]
31. Tlili, G.; Mesguich, C.; Gaye, D.; Tabarin, A.; Haissaguerre, M.; Hindié, E. Dual-tracer ^{99m}Tc-sestamibi/123I imaging in primary hyperparathyroidism. *Q. J. Nucl. Med. Mol. Imaging* **2023**, *67*, 114–121. [CrossRef] [PubMed]

32. Benderradji, H.; Beron, A.; Wémeau, J.L.; Carnaille, B.; Delcroix, L.; Do Cao, C.; Baillet, C.; Huglo, D.; Lion, G.; Boury, S.; et al. Quantitative dual isotope [123]iodine/^{99m}Tc-MIBI scintigraphy: A new approach to rule out malignancy in thyroid nodules. *Ann. Endocrinol.* **2021**, *82*, 83–91. [CrossRef]

33. Thuillier, P.; Benisvy, D.; Ansquer, C.; Corvilain, B.; Mirallié, E.; Taïeb, D.; Borson-Chazot, F.; Lussey-Lepoutre, C. SFE-AFCE-SFMN 2022 Consensus on the management of thyroid nodules: What is the role of functional imaging and isotopic treatment? *Ann. Endocrinol.* **2022**, *83*, 401–406. [CrossRef]

34. Peng, Y.; Pan, G.; Zhao, B.; Zuo, C.; Wang, Y.; Chen, R. Incremental value of ^{99m}Tc-MIBI single-photon emission computed tomography/computed tomography fusion imaging for the diagnosis of secondary hyperparathyroidism. *Nucl. Med. Commun.* **2023**, *44*, 767–771. [CrossRef] [PubMed]

35. Sankaran, S.J.; Davidson, J. Diagnosis and localization of parathyroid adenomas using 16-slice SPECT/CT: A clinicopathological correlation. *J. Med. Imaging Radiat. Oncol.* **2022**, *66*, 618–622. [CrossRef]

36. Guo, Y.H.; Huang, J.W.; Wang, Y.; Lu, R.; Yang, M.F. Value of ^{99m}Tc-MIBI SPECT/CT in the localization of recurrent lesions in patients with suspected recurrent parathyroid carcinoma. *Nucl. Med. Commun.* **2023**, *44*, 18–26. [CrossRef] [PubMed]

37. Blanco-Saiz, I.; Goñi-Gironés, E.; Ribelles-Segura, M.J.; Salvador-Egea, P.; Díaz-Tobarra, M.; Camarero-Salazar, A.; Rudic-Chipe, N.; Saura-López, I.; Alomar-Casanovas, A.; Rabines-Juárez, A.; et al. Preoperative parathyroid localization. Relevance of MIBI SPECT-CT in adverse scenarios. *Endocrinol. Diabetes Nutr.* **2023**, *70* (Suppl. S2), 35–44. [CrossRef]

38. Musumeci, M.; Pereira, L.V.; San Miguel, L.; Cianciarelli, C.; Vazquez, E.C.; Mollerach, A.M.; Arma, I.J.; Hume, I.; Galich, A.M.; Collaud, C. Normocalcemic primary hyperparathyroidism: [99m]Tc SestaMibi SPECT/CT results compare with hypercalcemic hyperparathyroidism. *Clin. Endocrinol.* **2022**, *96*, 831–836. [CrossRef]

39. Guenette, J.P. Opportunistic CT Screening for Parathyroid Adenoma. *Acad. Radiol.* **2023**, *30*, 891–892. [CrossRef]

40. Lincoln, C.M.M. Editorial Comment: Value-Added Assessment for Parathyroid Adenomas on Routine CT Examinations. *AJR Am. J. Roentgenol.* **2023**, *221*, 226. [CrossRef]

41. Sharma, A.; Patil, V.; Sarathi, V.; Purandare, N.; Hira, P.; Memon, S.; Jadhav, S.S.; Karlekar, M.; Lila, A.R.; Bandgar, T. Dual-phase computed tomography for localization of parathyroid lesions in children and adolescents with primary hyperparathyroidism. *Ann. Endocrinol.* **2023**, *84*, 446–453. [CrossRef] [PubMed]

42. Guo, M.; Lustig, D.B.; Lee, D.; Manhas, N.; Wiseman, S.M. Use of dual energy computed tomography versus conventional techniques for preoperative localization in primary hyperparathyroidism: Effect of preoperative calcium and parathyroid hormone levels. *Am. J. Surg.* **2023**, *225*, 852–856. [CrossRef] [PubMed]

43. Dugonjić, S.; Šišić, M.; Radulović, M.; Ajdinović, B. Positive ^{99m}Tc-MIBI and the subtraction parathyroid scan are related to intact parathyroid hormone but not to total plasma calcium in primary hyperparathyroidism. *Hell. J. Nucl. Med.* **2017**, *20*, 46–50.

44. Huynh, K.A.; MacFarlane, J.; Newman, C.; Gillett, D.; Das, T.; Scoffings, D.; Cheow, H.K.; Moyle, P.; Koulouri, O.; Harper, I.; et al. Diagnostic utility of ^{11}C-methionine PET/CT in primary hyperparathyroidism in a UK cohort: A single-centre experience and literature review. *Clin. Endocrinol.* **2023**, *99*, 233–245. [CrossRef] [PubMed]

45. Koumakis, E.; Gauthé, M.; Martinino, A.; Sindayigaya, R.; Delbot, T.; Wartski, M.; Clerc, J.; Roux, C.; Borderie, D.; Cochand-Priollet, B.; et al. FCH-PET/CT in Primary Hyperparathyroidism with Discordant/Negative MIBI Scintigraphy and Ultrasonography. *J. Clin. Endocrinol. Metab.* **2023**, *108*, 1958–1967. [CrossRef] [PubMed]

46. Saha, S.; Vierkant, R.A.; Johnson, G.B.; Parvinian, A.; Wermers, R.A.; Foster, T.; McKenzie, T.; Dy, B.; Lyden, M. C[11] choline PET/CT succeeds when conventional imaging for primary hyperparathyroidism fails. *Surgery* **2023**, *173*, 117–123. [CrossRef] [PubMed]

47. Stanciu, M.; Boicean, L.C.; Popa, F.L. The role of combined techniques of scintigraphy and SPECT/CT in the diagnosis of primary hyperparathyroidism: A case report. *Medicine* **2019**, *98*, e14154. [CrossRef] [PubMed]

48. Vestergaard, S.; Gerke, O.; Bay, M.; Madsen, A.R.; Stilgren, L.; Ejersted, C.; Rewers, K.I.; Jakobsen, N.; Asmussen, J.T.; Braad, P.E.; et al. Head-to-Head Comparison of Tc-99m-sestamibi SPECT/CT and C-11-L-Methionin PET/CT in Parathyroid Scanning Before Operation for Primary Hyperparathyroidism. *Mol. Imaging Biol.* **2023**, *25*, 720–726. [CrossRef]

49. Aphale, R.; Damle, N.; Chumber, S.; Khan, M.; Khadgawat, R.; Dharmashaktu, Y.; Agarwal, S.; Bal, C. Impact of Fluoro-Choline PET/CT in Reduction in Failed Parathyroid Localization in Primary Hyperparathyroidism. *World J. Surg.* **2023**, *47*, 1231–1237. [CrossRef]

50. Boudousq, V.; Guignard, N.; Gilly, O.; Chambert, B.; Mamou, A.; Moranne, O.; Zemmour, M.; Sharara, H.; Lallemant, B. Diagnostic Performance of Cervical Ultrasound, ^{99m}Tc-Sestamibi Scintigraphy, and Contrast-Enhanced ^{18}F-Fluorocholine PET in Primary Hyperparathyroidism. *J. Nucl. Med.* **2022**, *63*, 1081–1086. [CrossRef]

51. Freesmeyer, M.; Müller, U.A.; Männel, M.; Mtuka-Pardon, G.; Seifert, P. Synchronous Metastatic Medullary Thyroid Carcinoma and Paraesophageal Parathyroid Adenoma Detected on ^{18}F-Ethylcholine PET/US Fusion Imaging. *Clin. Nucl. Med.* **2022**, *47*, 977–979. [CrossRef] [PubMed]

52. Liu, Y.; Dang, Y.; Huo, L.; Hu, Y.; Wang, O.; Liu, H.; Chang, X.; Liu, Y.; Xing, X.; Li, F.; et al. Preoperative Localization of Adenomas in Primary Hyperparathyroidism: The Value of ^{11}C-Choline PET/CT in Patients with Negative or Discordant Findings on Ultrasonography and ^{99m}Tc-Sestamibi SPECT/CT. *J. Nucl. Med.* **2020**, *61*, 584–589. [CrossRef] [PubMed]

53. Noltes, M.E.; Rotstein, L.; Eskander, A.; Kluijfhout, W.P.; Bongers, P.; Brouwers, A.H.; Kruijff, S.; Metser, U.; Pasternak, J.D.; Veit-Haibach, P. 18F-fluorocholine PET/MRI versus ultrasound and sestamibi for the localization of parathyroid adenomas. *Langenbecks Arch. Surg.* **2023**, *408*, 155. [CrossRef] [PubMed]

54. Talbot, J.N.; Périé, S.; Tassart, M.; Delbot, T.; Aveline, C.; Zhang-Yin, J.; Kerrou, K.; Gaujoux, S.; Wagner, I.; Bennis, M.; et al. 18F-fluorocholine PET/CT detects parathyroid gland hyperplasia as well as adenoma: 401 PET/CTs in one center. *Q. J. Nucl. Med. Mol. Imaging* **2023**, *67*, 96–113. [CrossRef] [PubMed]

55. Klein, P.; Alsleibi, S.; Cohen, O.; Ilany, J.; Hemi, R.; Barhod, E.; Vered, I.; Winder, O.; Avior, G.; Tripto-Shklonik, L. Parathyroid fine-needle aspiration with parathyroid hormone washout as a preoperative localisation of parathyroid adenoma—A retrospective study. *Clin. Endocrinol.* **2023**, *99*, 246–252. [CrossRef] [PubMed]

56. Prades, J.M.; Lelonge, Y.; Farizon, B.; Chatard, S.; Prevot-Bitot, N.; Gavid, M. Positive predictive values of ultrasound-guided fine-needle aspiration with parathyroid hormone assay and Tc-99m sestamibi scintigraphy in sporadic primary hyperparathyroidism. *Eur. Ann. Otorhinolaryngol. Head. Neck Dis.* **2023**, *140*, 3–7. [CrossRef]

57. Amjad, W.; Trerotola, S.O.; Fraker, D.L.; Wachtel, H. Tricks of the trade: Techniques for preoperative localization in reoperative parathyroidectomy. *Am. J. Surg.* **2023**, *226*, 207–212. [CrossRef]

58. Hiramitsu, T.; Hasegawa, Y.; Futamura, K.; Okada, M.; Goto, N.; Narumi, S.; Watarai, Y.; Tominaga, Y.; Ichimori, T. Intraoperative intact parathyroid hormone monitoring and frozen section diagnosis are essential for successful parathyroidectomy in secondary hyperparathyroidism. *Front. Med.* **2022**, *9*, 1007887. [CrossRef]

59. Akgün, I.E.; Ünlü, M.T.; Aygun, N.; Kostek, M.; Uludag, M. Contribution of intraoperative parathyroid hormone monitoring to the surgical success in minimal invasive parathyroidectomy. *Front. Surg.* **2022**, *9*, 1024350. [CrossRef]

60. Habib, A.; Molena, E.; Snowden, C.; England, J. Efficacy of surgeon-performed, intra-operative ultrasound scan for localisation of parathyroid adenomas in patients with primary hyperparathyroidism. *J. Laryngol. Otol.* **2023**, *137*, 910–913. [CrossRef]

61. Girotti, P.N.C.; Gassner, J.; Hodja, V.; Tschann, P.; Königsrainer, I. "Before skin incision" high-resolution ultrasound in primary hyperparathyroidism: A new imaging tool for surgeons? *Langenbecks Arch. Surg.* **2022**, *407*, 3643–3649. [CrossRef] [PubMed]

62. Choi, J.H.; Jayaram, A.; Bresnahan, E.; Pletcher, E.; Steinmetz, D.; Owen, R.; Inabnet, W., III; Fernandez-Ranvier, G.; Taye, A. The Role of Surgeon-Performed Office and Preincision Ultrasounds in Localization of Parathyroid Adenomas in Primary Hyperparathyroidism. *Endocr. Pract.* **2022**, *28*, 660–666. [CrossRef] [PubMed]

63. Ishii, S.; Sugawara, S.; Yaginuma, Y.; Kobiyama, H.; Hiruta, M.; Watanabe, H.; Yamakuni, R.; Hakozaki, M.; Fujimaki, H.; Ito, H. Causes of false negatives in technetium-^{99m}methoxyisobutylisonitrile scintigraphy for hyperparathyroidism: Influence of size and cysts in parathyroid lesions. *Ann. Nucl. Med.* **2020**, *34*, 892–898. [CrossRef] [PubMed]

64. Stanciu, M.; Ristea, R.P.; Popescu, M.; Vasile, C.M.; Popa, F.L. Thyroid Carcinoma Showing Thymus-like Differentiation (CASTLE): A Case Report. *Life* **2022**, *12*, 1314. [CrossRef] [PubMed]

65. Saowapa, S.; Chamroonrat, W.; Suvikapakornkul, R.; Sriphrapradang, C. Incidental breast lesion detected by technetium-99m sestamibi scintigraphy in a patient with primary hyperparathyroidism. *World J. Nucl. Med.* **2019**, *19*, 69–71. [CrossRef]

66. Dumitru, N.; Ghemigian, A.; Carsote, M.; Albu, S.E.; Terzea, D.; Valea, A. Thyroid nodules after initial evaluation by primary health care practitioners: An ultrasound pictorial essay. *Arch. Balk. Med. Union* **2016**, *51*, 434–438.

67. Stanciu, M.; Zaharie, I.S.; Bera, L.G.; Cioca, G. Correlations between the presence of Hürthle cells and cytomorphological features of fine-needle aspiration biopsy in thyroid nodules. *Acta Endocrinol.* **2016**, *12*, 485–490. [CrossRef]

68. Lee, J.Y.; Song, H.S.; Choi, J.H.; Hyun, C.L.; Lee, S.A.; Choi, J.H.; Lee, S. Detecting Synchronous Thyroid Adenoma and False-Positive Findings on Technetium-99m MIBI Single Photon-Emission Computed Tomography/Computed Tomography. *Diagnostics* **2019**, *9*, 57. [CrossRef]

69. Paillahueque, G.; Massardo, T.; Barberán, M.; Ocares, G.; Gallegos, I.; Toro, L.; Araya, A.V. False negative spect parathyroid scintigraphy with sestamibi in patients with primary hyperparathyroidism. *Rev. Med. Chil.* **2017**, *145*, 1021–1027. [CrossRef]

70. Carral, F.; Jiménez, A.I.; Tomé, M.; Álvarez, J.; Díez, A.; Partida, F.; Ayala, C. Factors associated with negative ^{99m}Tc-MIBI scanning in patients with primary hyperparathyroidism. *Rev. Esp. Med. Nucl. Imagen Mol.* **2021**, *40*, 222–228. [CrossRef]

71. Filser, B.; Uslar, V.; Weyhe, D.; Tabriz, N. Predictors of adenoma size and location in primary hyperparathyroidism. *Langenbecks Arch. Surg.* **2021**, *406*, 1607–1614. [CrossRef] [PubMed]

72. Buicko, J.L.; Kichler, K.M.; Amundson, J.R.; Scurci, S.; Kozol, R.A. The Sestamibi Paradox: Improving Intraoperative Localization of Parathyroid Adenomas. *Am. Surg.* **2017**, *83*, 832–835. [CrossRef] [PubMed]

73. Palot Manzil, F.F.; Eichhorn, J.; Vattoth, S. Synchronous Ectopic Thyroid Gland and Ectopic Parathyroid Adenoma on ^{99m}Tc-Sestamibi Scintigraphy and Correlative Imaging. *J. Nucl. Med. Technol.* **2023**, *51*, 263–264. [CrossRef] [PubMed]

74. Zhang, W.; Liu, F.; Chen, K.; Wang, Y.; Dou, J.; Mu, Y.; Lyu, Z.; Zang, L. Case report: Coexistence of primary hyperparathyroidism with giant toxic nodular goiter. *BMC Endocr. Disord.* **2022**, *22*, 200. [CrossRef]

75. Cansu, G.B.; Taşkıran, B.; Bahçeci, T. Thyroid Hemiagenesis Associated with Graves' Disease: A Case Report and Review of the Literature. *Acta Endocrinol.* **2017**, *13*, 342–348. [CrossRef]

76. Lesi, O.K.; Thapar, A.; Appaiah, N.N.B.; Iqbal, M.R.; Kumar, S.; Maharaj, D.; Saad Abdalla Al-Zawi, A.; Dindyal, S. Thyroid Hemiagenesis: Narrative Review and Clinical Implications. *Cureus* **2022**, *14*, e22401. [CrossRef]

77. Ren, Y.; Jiang, G.; Meng, Y.; Chen, J.; Liu, J. ^{99m}Tc-pertechnetate thyroid static scintigraphy unexpectedly revealed ectopic gastric mucosa of upper esophagus. *Hell. J. Nucl. Med.* **2023**, *26*, 157–159.

78. Giovanella, L.; Petranović Ovčariček, P. Functional and molecular thyroid imaging. *Q. J. Nucl. Med. Mol. Imaging* **2022**, *66*, 86–92. [CrossRef]

79. Campennì, A.; Giovanella, L.; Siracusa, M.; Alibrandi, A.; Pignata, S.A.; Giovinazzo, S.; Trimarchi, F.; Ruggeri, R.M.; Baldari, S.; Trimboli, P.; et al. (99m)Tc-Methoxy-Isobutyl-Isonitrile Scintigraphy Is a Useful Tool for Assessing the Risk of Malignancy in Thyroid Nodules with Indeterminate Fine-Needle Cytology. *Thyroid* **2016**, *26*, 1101–1109. [CrossRef]
80. Campennì, A.; Siracusa, M.; Ruggeri, R.M.; Laudicella, R.; Pignata, S.A.; Baldari, S.; Giovanella, L. Differentiating malignant from benign thyroid nodules with indeterminate cytology by ^{99m}Tc-MIBI scan: A new quantitative method for improving diagnostic accuracy. *Sci. Rep.* **2017**, *7*, 6147. [CrossRef]
81. Arabi, M.; Zamani, H.; Soltanabadi, M.; Kalhor, L. [^{99m}Tc]MIBI scintigraphy in a patient with thyroid follicular neoplasm: A case report and review of literature. *Nucl. Med. Rev. Cent. East. Eur.* **2021**, *24*, 118–119. [CrossRef] [PubMed]
82. Schenke, S.A.; Görges, R.; Seifert, P.; Zimny, M.; Kreissl, M.C. Update on diagnosis and treatment of hyperthyroidism: Ultrasonography and functional imaging. *Q. J. Nucl. Med. Mol. Imaging* **2021**, *65*, 102–112. [CrossRef] [PubMed]

Article

Surgical Outcomes of Pancreatic Solid Pseudopapillary Neoplasm: Experiences of 24 Patients in a Single Institute

Peng-Yu Ku [1], Shao-Bin Cheng [2], Yi-Ju Chen [1], Chia-Yu Lai [2], Hsiao-Tien Liu [2] and Wei-Hsin Chen [1,*]

[1] Division of General Surgery, Department of Surgery, Taichung Veterans General Hospital, Taichung 40705, Taiwan; lukephil6@gmail.com (P.-Y.K.)
[2] Division of General Surgery, Department of Surgery, Taichung Tzu Chi Hospital, Taichung 427213, Taiwan
* Correspondence: maxchen0127@yahoo.com.tw

Abstract: *Background and Objectives*: The pancreatic solid pseudopapillary neoplasm (SPN), a rare tumor predominantly affecting young women, has seen an increased incidence due to improved imaging and epidemiological knowledge. This study aimed to understand the outcomes of different interventions, possible complications, and associated risk factors. *Materials and Methods*: This study retrospectively analyzed 24 patients who underwent pancreatic surgery for SPNs between September 1998 and July 2020. *Results*: Surgical intervention, typically required for symptomatic cases or pathological confirmation, yielded favorable outcomes with a 5-year survival rate of up to 97%. Despite challenges in standardizing preoperative evaluation and follow-up protocols, aggressive complete resection showed promising long-term survival and good oncological outcomes. Notably, no significant differences were found between conventional and minimally invasive (MI) surgery in perioperative outcomes. Histopathological correlations were lacking in prognosis and locations. Among the patients, one developed diffuse liver metastases 41 months postoperatively but responded well to chemotherapy and transcatheter arterial chemoembolization, with disease stability observed at 159 postoperative months. Another patient developed nonalcoholic steatohepatitis after surgery and underwent liver transplantation, succumbing to poor medication adherence 115 months after surgery. *Conclusions*: These findings underscore the importance of surgical intervention in managing SPNs and suggest the MI approach as a viable option with comparable outcomes to conventional surgery.

Keywords: solid pseudopapillary neoplasm of the pancreas; surgical outcomes; pancreatic endocrine function; pancreatic exocrine function

Citation: Ku, P.-Y.; Cheng, S.-B.; Chen, Y.-J.; Lai, C.-Y.; Liu, H.-T.; Chen, W.-H. Surgical Outcomes of Pancreatic Solid Pseudopapillary Neoplasm: Experiences of 24 Patients in a Single Institute. *Medicina* **2024**, *60*, 889. https://doi.org/10.3390/medicina60060889

Academic Editor: Jin Wook Yi

Received: 6 May 2024
Revised: 22 May 2024
Accepted: 25 May 2024
Published: 28 May 2024

1. Introduction

Pancreatic solid pseudopapillary neoplasm (SPN) was first described by Frantz in 1959, based on the pathologic examination of three patients [1]. In 1970, Hamoudi and colleagues expanded the literature by adding an additional patient and providing detailed electron microscopic descriptions of the tumor [2]. SPNs are characterized by their gross and histologic appearance, which includes discohesive polygonal cells surrounding delicate blood vessels and forming a solid mass, often with cystic degeneration and intracystic hemorrhage [3]. The neoplastic cells exhibit uniform nuclei, finely stippled chromatin, and nuclear grooves, along with eosinophilic globules, foam cells, and cholesterol clefts [3].

SPNs have also been referred to as solid and papillary tumors, papillary cystic tumors, solid cystic tumors, Hamoudi tumors, or Frantz tumors [4]. Despite these varying names, SPN is the preferred terminology [3]. SPNs are rare, comprising only 1–2% of all pancreatic tumors [4,5]. In 1996, the World Health Organization classified pancreatic SPNs as a borderline malignancy of the exocrine pancreas [6]. The incidence of pancreatic SPNs has increased over the past two decades due to improved imaging techniques and a better epidemiological understanding of the disease [7]. SPNs are most commonly diagnosed in women aged 30–40 years and have a favorable prognosis, with a 5-year survival rate of up to 97% [8,9]. Before the operation, imaging studies were conducted to identify the SPN, utilizing sonography (US), computed tomography (CT), and magnetic resonance imaging (MRI). About one-third of pancreatic SPN cases are located in the pancreatic tail, and another third in the pancreatic head [10]. The benefits of the increasing preoperative pathology diagnosis of SPN using endoscopic ultrasonography (EUS) have yet to be fully established [11].

Surgical intervention for SPNs was first mentioned by Sanfey and associates in 1983 and is often indicated based on tumor-related symptoms or for pathological diagnosis [12]. Although SPNs are usually localized, 10% to 15% of patients may develop metastases [13,14]. These metastases are often resectable, and complete removal is associated with long-term survival [14,15]. Recent studies have compared the outcomes of conventional distal pancreatectomies with minimally invasive (MI) pancreatic surgeries, including laparoscopic and robot-assisted approaches, which have become increasingly favored [16]. Post-operative oncological outcomes are generally positive, with only 15% of SPN cases progressing to distant metastasis, typically in the liver or peritoneum [10,14].

In our study, we detail our experience in treating pancreatic SPNs, including pre-operative imaging results and pathology diagnoses, procedure comparisons, surgical outcomes, and complications in patients with pathologically confirmed pancreatic SPNs.

2. Materials and Methods

2.1. Study Group

We collected the data of patients who underwent pancreatic surgery between September 1998 and July 2020, with a follow-up period ending in March 2023. The patients' medical records were retrospectively reviewed to identify those who underwent either conventional or MI surgery for pathologically confirmed SPNs in our hospital, a tertiary referral center in Central Taiwan; procedures included pancreaticoduodenectomy (PD), pylorus-preserving pancreaticoduodenectomy (PPPD), partial pancreatectomy (PP), central pancreatectomy (CP), and distal pancreatectomy (DP). A total of 1010 patients underwent pancreatic-related operations, of which 692 underwent PD, and 318 underwent either DP or PP. In total, 25 patients were enrolled in this study according to the pathology report confirming SPN, but 1 patient was excluded due to insufficient data (Figure 1).

2.2. Data Collection

We reviewed data, including clinical characteristics, such as sex, age, BMI, and tumor markers; pre-operative imaging studies; and EUS reports. The results of fine needle EUS biopsy for preoperative pathology were analyzed to compare its efficacy with that of preoperative imaging studies.

The surgical outcomes and pathology data analyzed included tumor location and size, status of resection margin, EBL, day of enteral feeding, day of drainage tube removal, and length of hospitalization. The postoperative outcomes were also collected and included complications, postoperative pancreatic fistula, DGE, pathology reports, and post-pancreatectomy diabetes.

2.3. Definition of Outcomes

Tumor size was measured and recorded from pathology reports. Blood loss of <50 mL in our hospital was recorded as minimal EBL for the statistical analysis. The amylase concentration in the drainage fluid was exanimated in all the patients, and pancreatic fistulas were graded as follows: biochemical leak, with no clinical impact; grade B, requiring a change in management or adjustment of the clinical approach; and grade C, requiring a major change in clinical management or deviation from the normal clinical approach, according to the Pancreatic Fistula Classification in the International Study Group on Pancreatic Fistula definition [17].

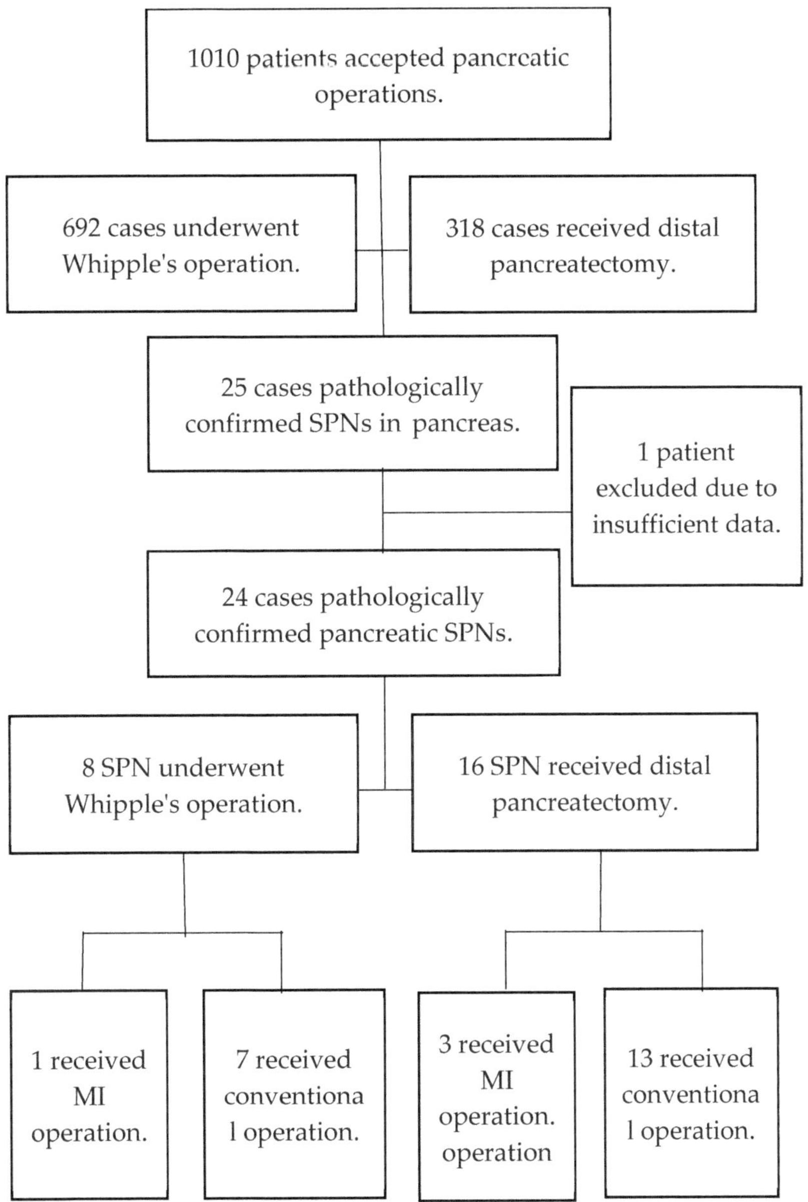

Figure 1. Patient selection flowchart. SPN solid pseudopapillary neoplasm. MI minimal invasive.

According to the International Study Group of Pancreatic Surgery definition, we also defined DGE according to the duration of nasogastric intubation: 4–7 postoperative days (PODs) as Grade A (mild); 8–14 PODs as Grade B (moderate); and >14 PODs as Grade C (severe) [18].

Post-pancreatectomy diabetes was defined according to American Diabetes Association 2022 guidelines as a fasting plasma sugar of >126 mg/dL or an HbA1c of >6.5% detected during the follow-up period [19].

2.4. Statistical Analysis

IBM SPSS version 22.0 (International Business Machines Corp, New York, NY, USA) was used for the statistical analyses. Categorical variables were expressed as percentages, frequencies, or medians with interquartile ranges. Chi-square tests or Mann–Whitney U tests were used to compare categorical data. The correlations between two variables were analyzed with the Pearson correlation coefficient.

2.5. Ethics Approval and Consent to Participate

The study adhered to the principles outlined in the Declaration of Helsinki and the International Conference on Harmonisation—Good Clinical Practice guidelines. The study was also reviewed and approved by the Institutional Review Board I & II of Taichung Veterans General Hospital (TCVGH-IRB No.: CE22251A) on 14/06/2022. The need for informed consent was waived by the Institutional Review Board I & II of Taichung Veterans General Hospital.

3. Results

3.1. λ General Results

Of the 24 patients included in this study, 21 were women, and 3 were men (women-to-men ratio of 7:1) (Table 1). The median age of patients was 34.5 (23.8–41.5) years, with a median body mass index (BMI) of 20.8 (18.9–22.4) kg/m^2.

Most of our patients were asymptomatic (11/24 cases, 46%), while some of the pancreatic SPNs presented with abdominal pain (8/24 cases, 29%) or palpable abdominal masses (4/24 cases, 17%).

Pancreatic SPNs were predominant in young women in all groups we classified, either by location or procedure (conventional and MI surgery) (Table 1). No statistical differences were identified in terms of age, sex, BMI, size, and tumor marker in comparisons of locations or procedures (Table 1).

Nevertheless, a trend was identified in the procedure comparison; the MI group had a younger age, compared with the conventional group (median, MI 23.5 [19.5–27.5] vs. conventional 38.0 [28.0–45.8] cm; $p = 0.052$).

3.2. Preoperative Diagnosis

All of our patients had preoperative radiologic findings in either US, CT, or MRI. Most showed typical features of SPN (Figures 2 and 3), such as oval, exophytic, and regular capsulated lesions with a mixed cystic and solid component but were almost entirely solid or cystic with thick walls [7]. Preoperative imaging studies included US (10/24 cases), CT (23/24 cases), and MRI (14/24 cases). In total, nine cases were diagnosed as SPNs based on preoperative imaging, four cases based on CT (17%), and five cases based on MRI (36%) (Table 1). The CT modalities are shown in Table 2; all tumor margins were clear in preoperative images. No hemorrhage, parenchyma atrophy, or invasion of adjacent vessels and abdominal organs was seen. More irregular shaping was seen in SPNs of the pancreatic head ($p = 0.037$). The cystic component was significantly more prominent in the SPN of the pancreatic tail ($p = 0.035$). Pancreas or bile duct dilation was mostly seen in the pancreatic head in our patient ($p < 0.001$).

Preoperative pathological EUS fine needle biopsies were performed in 11 patients, with only 2 cases being preoperatively diagnosed as pancreatic SPNs (Table 1).

Of these 24 patients, 8 were diagnosed with pancreatic SPNs located at the pancreatic head, and 16 were diagnosed with pancreatic SPNs located at the distal pancreas (tail, 9 cases, 37.5%; body and tail, 2 cases 8.3%; body, 5 cases, 20.8%). We included the body and tail group and the tail group into the "body to tail group" for analysis between locations in Table 1.

Table 1. Patient characteristics and surgical outcomes between different locations (n = 24).

	Total (n = 24)		Head (n = 8)		Body (n = 5)		* Body to Tail (n = 11)		*p*-Value
Age	34.5	(23.8–41.5)	30.5	(18.0–45.8)	28	(21.5–56)	38	(28.0–40.0)	0.596
Sex									1.000
Female	21	(87.5%)	7	(87.5%)	4	(80%)	10	(91%)	
Male	3	(12.5%)	1	(12.5%)	1	(20%)	1	(9%)	
BMI	20.8	(18.9–22.4)	19.0	(18.3–23.4)	21.1	(18.8–21.9)	21.1	(20.4–22.9)	0.561
Size	7.6	(4.1–9.2)	6.0	(4.1–9.9)	3.2	(1.6–7.5)	9.0	(7.5–9.3)	0.149
CEA	1.5	(1.1–3.3)	1.6	(1.1–3.3)	1.5	(1.1–3.1)	1.5	(1.1–3.4)	0.992
CA19-9	12.8	(5.7–29.1)	10.8	(5.0–19.4)	26.8	(8.3–457.8)	12.8	(5.3–29.9)	0.467
Preop diagnosis									
US	10	(41.7%)	3	(37.5%)	2	(40%)	5	(46%)	1.000
CT	23	(95.8%)	8	(100%)	5	(100%)	10	(91%)	0.540
SPN in CT	4/23	(17.4%)	1/8	(12.5%)	0/5	(0%)	3/10	(30%)	0.368
MRI	14	(58.3%)	7	(87.5%)	4	(80%)	3	(27%)	0.021 *
SPN in MRI	5/14	(35.7%)	2/7	(28.6%)	2/4	(50%)	1/3	(33%)	0.790
EUS	11	(45.8%)	5	(62.5%)	4	(80%)	2	(18%)	0.047 *
SPN in EUS	2/11	(18.2%)	1/5	(20.0%)	1/4	(25%)	0	(0%)	1.000
Operations									
Conventional	20	(83.3%)	7	(87.5%)	3	(60%)	10	(89%)	
PD or PPPD	7	(29%)	7	(0%)	0	(0%)	0	(0%)	
DP	13	(48%)	0	(0%)	3	(60%)	10	(89%)	
CP	0	(0%)	0	(0%)	0	(0%)	0	(0%)	
MI	4	(16.7%)	1	(12.5%)	2	(40%)	1	(11%)	
PD or PPPD	1	(4%)	1	(12.5%)	0	(0%)	0	(0%)	
DP	2	(8%)	0	(0%)	1	(20%)	1	(11%)	
CP	1	(4%)	0	(0%)	1	(20%)	0	(0%)	
Complications									
EBL, mL	200	(50–665)	380	(212.5–803.8)	50	(50–125)	260	(50–738)	0.028 *
EF, days	4	(3–5)	5	(4–8)	4	(2–5)	3.5	(3–4)	0.017 *
RD, days	11	(6–22)	18	(7–20)	13	(5–25.5)	7	(6–26)	0.980
LoH, days	10	(7–21)	21	(12–27)	8	(7–11.5)	7.5	(7–13)	0.010 *
F/U, months	65.5	(34.5–116.5)	99	(61.5–139)	41	(9–108)	52	(26–99)	0.248
Leakage, ISPGF									0.432
N	10	(43.5%)	4	(50.0%)	1	(20.0%)	5	(50%)	
A	6	(26.1%)	3	(37.5%)	2	(40.0%)	1	(10%)	
B	7	(30.4%)	1	(12.5%)	2	(40.0%)	4	(40%)	
DGE, ISPGS									0.083
N	18	(81.8%)	4	(57.1%)	4	(80.0%)	10	(100%)	
A	2	(9.1%)	1	(14.3%)	1	(20.0%)	0	(0%)	
B	2	(9.1%)	2	(28.6%)	0	(0%)	0	(0%)	

* Chi-square test or Mann–Whitney U test, median (IQR); * *p* < 0.05. Body to tail group included the body and tail and the tail group. DP distal pancreatectomy, CP central pancreatectomy, PD Whipple operation, PPPD pylorus-preserving pancreaticoduodenectomy, MI minimal invasive, BMI body mass index, CEA carcinoembryonic antigen, CA19-9 carbohydrate antigen 19-9, US ultrasonography, CT computed tomography, SPN solid pseudopapillary neoplasm, MRI magnetic resonance imaging, EUS endoscopic ultrasonography, EBL estimated blood loss, RD remove drainage, LoH length of hospitalization, SPN solid pseudopapillary neoplasm, ISGPF International Study Group on Pancreatic Fistula, DGE delayed gastric emptying, ISGPS International Study Group of Pancreatic Surgery.

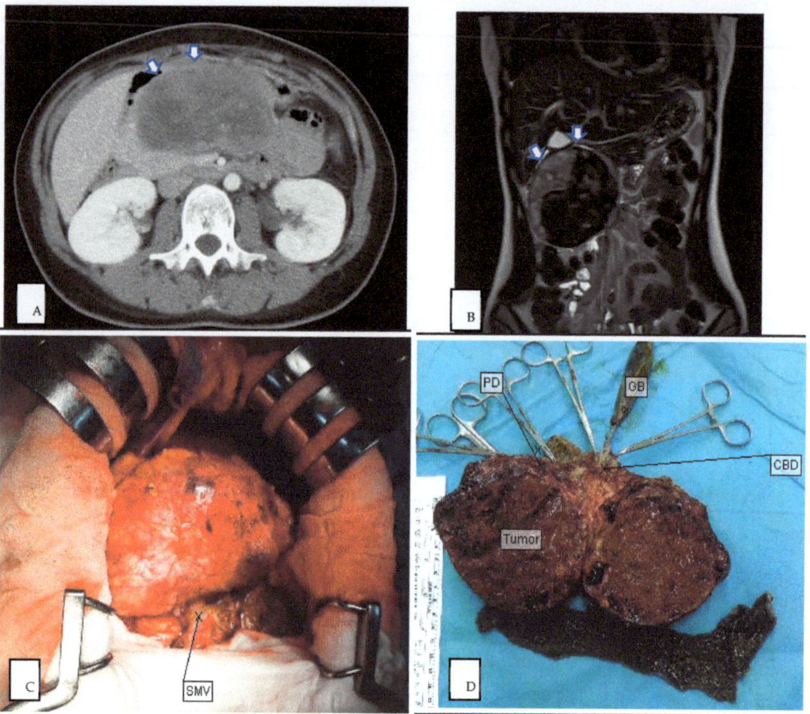

Figure 2. (**A**) Preoperative CT image of SPN in pancreatic head (axial plane, venous phase). (**B**) Preoperative MRI image of SPN in pancreatic tail (coronal plane, T2 weighted). (**C**) Intraoperative photography of SPN in pancreatic head. (**D**) Cross section of intraoperative specimen in pancreatic head, showing a tumor composed of mixed cystic and solid components with hemorrhagic areas. White arrow demarcates region of pancreatic tail mass. PD pancreaticoduodenectomy, GB gallbladder, CBD common bile duct.

3.3. Postoperative Outcomes and Complications

Among the patients with distal pancreatic SPNs, 13 underwent DP with splenectomy, 1 underwent DP without splenectomy, 1 underwent single port laparoscopic DP with splenectomy, 1 underwent robotic (Xi)-assisted CP and gastropancreaticostomy without splenectomy, and 1 underwent laparoscopic DP with splenectomy (Table 3). The median tumor size in the pathologic report was 7.6 (4.1–9.2) cm. Furthermore, a trend was identified that patients in the MI group had smaller tumor sizes (median, MI 3.1 [1.7–7.6] vs. conventional 7.8 [5.0–9.4] cm; $p = 0.052$).

No margin involvement was present in the pathologic reports of the enrolled patients. The immunohistochemistry (IHC) stains were mostly detected in CD10 (14 cases, 100% in CD10 stain cases), CD56 (7 cases, 100% in CD56 stain cases), and B-catenin (12 cases, 100% in B-catenin stain cases) (Table 4). Little positive immunolabeling was found in synaptophysin, chromogranin, vimentin, NSE, alpha-1-antichymotrypsin, and CyclinD1 but no strong correlation was detected. No histopathological correlation was found in prognoses or locations (Table 4).

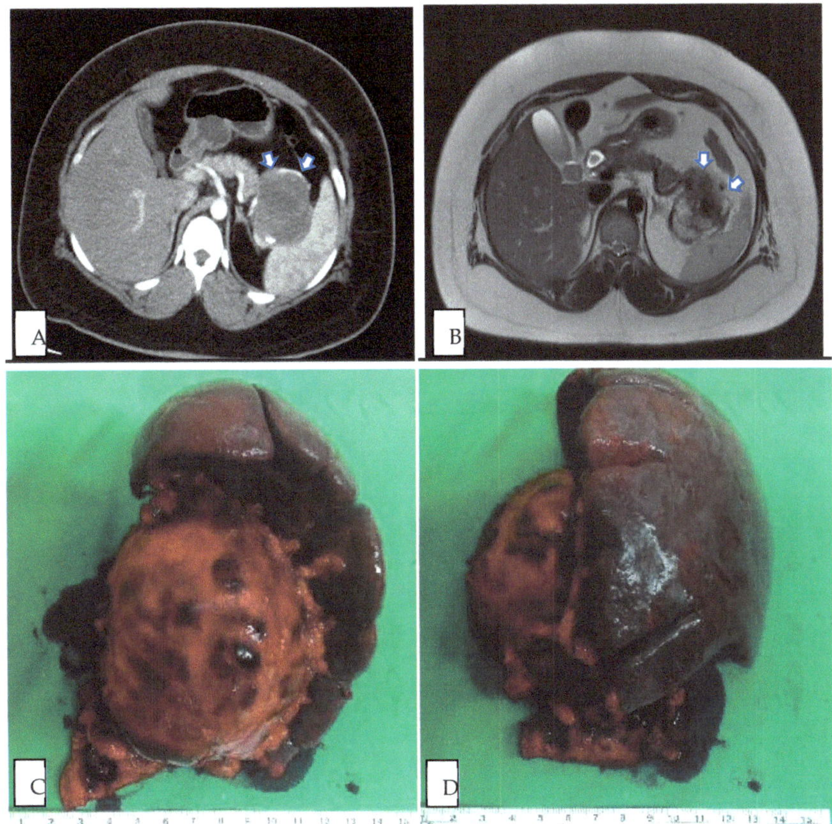

Figure 3. (**A**) Preoperative CT image of SPN in pancreatic tail (axial plane, venous phase). (**B**) Preoperative MRI image of SPN in pancreatic tail (axial plane, T2 weighted). (**C**) Intraoperative specimen of DP with splenectomy from SPN in pancreatic tail, ventral view. (**D**) Intraoperative specimen of DP with splenectomy from SPN in pancreatic tail, dorsal view. White arrow demarcates region of pancreatic tail mass.

Table 2. Computed tomography characteristics in our patients.

	Total (n = 24)		Head (n = 8)		Body (n = 5)		* Body to Tail (n = 11)		*p*-Value
Shape									0.037 *
Oval	20	(87.0%)	5	(62.5%)	5	(100.0%)	10	(100.0%)	
Irregular	3	(13.0%)	3	(37.5%)	0	(0.0%)	0	(0.0%)	
Ratio of solid-to-cystic components									0.035 *
Solid	9	(39.1%)	4	(50.0%)	3	(60.0%)	2	(20.0%)	
Half	6	(26.1%)	4	(50.0%)	0	(0.0%)	2	(20.0%)	
Cystic	8	(34.8%)	0	(0.0%)	2	(40.0%)	6	(60.0%)	
Capsule complete									0.308
N	2	(8.7%)	1	(12.5%)	1	(20.0%)	0	(0.0%)	
Y	21	(91.3%)	7	(87.5%)	4	(80.0%)	10	(100.0%)	
Margin clear									--
N	0	(0.0%)	0	(0.0%)	0	(0.0%)	0	(0.0%)	
Y	23	(100.0%)	8	(100.0%)	5	(100.0%)	10	(100.0%)	

Table 2. *Cont.*

	Total (n = 24)		Head (n = 8)		Body (n = 5)		* Body to Tail (n = 11)		*p*-Value
Growth pattern									0.082
Exophytic	20	(87.0%)	7	(87.5%)	3	(60.0%)	10	(100.0%)	
Intra	3	(13.0%)	1	(12.5%)	2	(40.0%)	0	(0.0%)	
Margin calcification									0.663
n	14	(60.9%)	6	(75.0%)	3	(60.0%)	5	(50.0%)	
y	9	(39.1%)	2	(25.0%)	2	(40.0%)	5	(50.0%)	
Hemorrhage									--
n	23	(100.0%)	8	(100.0%)	5	(100.0%)	10	(100.0%)	
y	0	(0.0%)	0	(0.0%)	0	(0.0%)	0	(0.0%)	
Compression of the main pancreatic duct and bile duct									0.565
n	22	(95.7%)	7	(87.5%)	5	(100.0%)	10	(100.0%)	
y	1	(4.3%)	1	(12.5%)	0	(0.0%)	0	(0.0%)	
Upstream pancreatic parenchymal atrophy									--
n	23	(100.0%)	8	(100.0%)	5	(100.0%)	10	(100.0%)	
y	0	(0.0%)	0	(0.0%)	0	(0.0%)	0	(0.0%)	
Invasion of adjacent vessels and abdominal organs									--
n	23	(100.0%)	8	(100.0%)	5	(100.0%)	10	(100.0%)	
y	0	(0.0%)	0	(0.0%)	0	(0.0%)	0	(0.0%)	
Pancreas or bile duct dilation									<0.001 **
n	15	(65.2%)	1	(12.5%)	4	(80.0%)	10	(100.0%)	
Dilate	8	(34.8%)	7	(87.5%)	1	(20.0%)	0	(0.0%)	

Fisher's exact test. * $p < 0.05$, ** $p < 0.01$. There was one patient without any pre-operative image study. Body to tail group included the body and tail and the tail group intra intrapancreatic.

Table 3. Procedure list.

	Procedure	Number
Conventional	PD	3
	PPPD	4
	DP (body) with splenectomy	10
	DP (body) with splenectomy and stomach wedge resection	1
	DP (body) with spleen preservation	1
	DP (tail) with splenectomy and radical nephrectomy	1
Minimally invasive	Robotic (Si) assisted PPPD	1
	Robotic (Xi) assisted CP and gastropancreaticostomy with spleen preservation	1
	Single port laparoscopic DP (tail) with spleen preservation	1
	Laparoscopic DP (Tail)	1

Si Da Vinci Si surgical system, Xi Da Vinci Xi surgical system, PD pancreaticoduodenectomy, PPPD Pylorus preserving pancreaticoduodenectomy, DP distal pancreatectomy, CP central pancreatectomy.

Three complications were registered in the records: bile leakage and hepato-jejunostomy stricture requiring endoscopic balloon dilation after PPPD, postoperative pseudocyst after PPPD requiring no surgical intervention, and postoperative pancreatic tail hematoma after DP (body) with splenectomy, which subsided under conservative treatment. The cases included in this study were followed up till March 2023, with a median follow-up of 65.5 (34.5–116.5) months. The longest follow-up observed was 229 months after DP, and the patient had disease stability at the last visit. Further, 16 patients were lost during follow-up, and 6 patients had stable disease statuses at the end of our study.

Table 4. Histopathological profiles of SPNs in our patients.

	Total (n = 24)		Head (n = 8)		Body (n = 5)		* Body to Tail (n = 11)		*p*-Value
B-catenin									1.000
NT	12	(50.0%)	4	(50.0%)	3	(60.0%)	5	(45.5%)	
P	12	(50.0%)	4	(50.0%)	2	(40.0%)	6	(54.5%)	
N	0	(0.0%)	0	(0.0%)	0	(0.0%)	0	(0.0%)	
CD10									1.000
NT	8	(33.3%)	3	(37.5%)	1	(20.0%)	4	(36.4%)	
P	16	(66.7%)	5	(62.5%)	4	(80.0%)	7	(63.6%)	
N	0	(0.0%)	0	(0.0%)	0	(0.0%)	0	(0.0%)	
CD56									1.000
NT	2	(8.7%)	1	(12.5%)	1	(20.0%)	0	(0.0%)	
P	21	(91.3%)	7	(87.5%)	4	(80.0%)	10	(100.0%)	
N	0	(0.0%)	0	(0.0%)	0	(0.0%)	0	(0.0%)	
Chromatin									0.310
NT	10	(41.7%)	4	(50.0%)	0	(0.0%)	6	(54.5%)	
P	6	(25.0%)	2	(25.0%)	2	(40.0%)	2	(18.2%)	
N	8	(33.3%)	2	(25.0%)	3	(60.0%)	3	(27.3%)	
Synaptophysin									0.153
NT	20	(83.3%)	7	(87.5%)	3	(60.0%)	10	(90.9%)	
P	3	(12.5%)	0	(0.0%)	2	(40.0%)	1	(9.1%)	
N	1	(4.2%)	1	(12.5%)	0	(0.0%)	0	(0.0%)	
Cyclin D1									0.803
NT	20	(83.3%)	6	(75.0%)	4	(80.0%)	10	(90.9%)	
P	3	(12.5%)	1	(12.5%)	1	(20.0%)	1	(9.1%)	
N	1	(4.2%)	1	(12.5%)	0	(0.0%)	0	(0.0%)	
Vimentin									1.000
NT	16	(66.7%)	6	(75.0%)	3	(60.0%)	7	(63.6%)	
P	8	(33.3%)	2	(25.0%)	2	(40.0%)	4	(36.4%)	
N	0	(0.0%)	0	(0.0%)	0	(0.0%)	0	(0.0%)	
PR									0.501
NT	16	(66.7%)	4	(50.0%)	4	(80.0%)	8	(72.7%)	
P	6	(25.0%)	2	(25.0%)	1	(20.0%)	3	(27.3%)	
N	2	(8.3%)	2	(25.0%)	0	(0.0%)	0	(0.0%)	

Fisher's exact test. * $p < 0.05$. Body to tail group included the body and tail and the tail group. NT not tested, P positive, N negative, SPN, solid pseudopapillary neoplasm.

A 38-year-old woman developed liver metastasis 41 months after undergoing PD, receiving chemotherapy once, and refusing subsequent doses due to severe side effects; she initiated palliative transcatheter arterial chemoembolization (TACE). Spleen metastasis and carcinomatosis were successively found, and she was lost during follow-up 10 years postoperatively.

A 9-year-old girl suffered from jaundice after PD, and nonalcoholic steatohepatitis (NASH) was diagnosed through liver biopsy 18 months after PD. She received living-related liver transplantation 70 months after the operation and passed away 115 months after PD due to graft failure related to psychological conditions and poor immunosuppressant adherence.

A 28-year-old woman, with a 7.6 cm tumor over the pancreatic tail, developed post pancreatectomy diabetes (1/27, 4.2%) in the follow-up period, after DP (body) with splenectomy.

During the follow-up period, two patient deaths occurred; one died after liver transplantation, as mentioned above, and the other died of colon cancer with liver and lung metastasis 12 months after receiving DP with splenectomy.

Statistically significant differences were identified between locations in terms of estimated blood loss (EBL), day of enteral feeding, and length of hospitalization in our study. No statistically significant differences were found between conventional and MI groups in terms of EBL, day of enteral feeding, day of drainage tube removal, length of hospitalization, follow-up period, post-operative pancreatic fistula grade, and delayed gastric emptying (DGE) grade in our study.

4. Discussion

In our single-center experience of pancreatic SPNs, only 12.5% of the patients were men in our population, which is consistent with another report, revealing that 10% of these tumors are diagnosed in men [5]. The median age of our patients was 38.0 (28.0–45.8) years in the conventional group and 23.5 (19.5–27.5) years in the MI group, without significant differences, as others have reported typically in the 2–3 decades of life [20]. The younger age in the MI group may be attributed to the progression of the health check concept in the modern world, the higher cosmetic demand in young populations, and better accessibility to image examinations.

Patients in our study underwent conventional surgery or MI surgery for SPNs, showing no statistically significant differences in terms of sex, age, BMI, and perioperative outcomes, which are similar to the findings of previous reports on pancreatic SPN management [21,22]. Most symptoms of our patients were nonspecific, and diagnosis was incidentally encountered through imaging examinations. Other symptoms included abdominal pain (eight cases, 29%), palpable mass lesions (four cases, 17%), and tarry stool (one case, 4.2%), which aligns with previous research [5,10,20,21]. Tumor markers, such as carbohydrate antigen (CA) 19-9 and carcinoembryonic antigen, were almost within the normal range, as previous studies have reported [4,10,13,21].

CT is the most commonly used preoperative imaging study and plays a significant role in the diagnosis of cystic lesions of the pancreas due to its cost-effectiveness in detecting and characterizing pancreatic SPNs, while MRI may allow for better identification of several tissue characteristics, such as hemorrhage, cystic degeneration, or the presence of a capsule [10,20,23,24]. If MRI reveals a well-marginated, encapsulated, solid, and cystic mass, with areas of hemorrhagic degeneration, a diagnosis of pancreatic SPN should be considered [23]. The role of EUS fine needle biopsy in pancreatic SPNs is not entirely established due to a lack of precise preoperative pathological diagnoses (diagnosis ratio: 2/11, 18.18% in our study). Furthermore, EUS fine needle biopsy can also lead to complications, such as rupture and peritoneal seeding, according to a previous study [11]. According to our experience, EUS fine needle biopsy was applied more frequently to exclude other malignant pancreatic lesions than to confirm pancreatic SPN diagnosis. Research has also been conducted on the non-invasive detection of pancreatic cancer utilizing fecal microbiota signatures in conjunction with serum levels of CA19-9 [25,26]. This method could potentially improve our preoperative differentiation between SPNs and pancreatic cancer.

The most common location of SPNs in our study was the pancreatic tail (nine cases, 37.5%), with other locations including the pancreatic head (eight cases, 33.3%), body (five cases, 21%), and body and tail (two cases, 8.3%), which is similar to those reported by Kang and Uğuz [14,27]. In image modalities, we found more irregular shapes seen in the pancreatic head ($p = 0.037$) and more cystic components in the pancreatic tail ($p = 0.035$), which were not well established in previous research. We may find better correlations in image modalities based on locations for better preoperative diagnoses with greater data collection. Pancreas or bile duct dilation was mostly seen in SPNs of the pancreatic head ($p < 0.001$), which may be related to the pancreas anatomy.

We also found an increasing incidence of SPN in the past 10 years in our patient group (10 patients before 2010 and 14 patients after 2010), which may be due to an increase in the health check-up concept, imaging studies, EUS fine needle aspiration, and pathology findings, as reported by Law et al. [9].

The statistical significance in postoperative outcome comparisons between locations may result from the different complexities of anatomy between PD (including PD and PPPD) and DP (including CP and partial pancreatectomy).

The smaller size of the MI surgery group (median, 3.1 [1.7–7.6] vs. 7.8 [5.0–9.4] cm) may be due to the advances in imaging modalities that lead to higher accuracy in the diagnosis of pancreatic SPN, as reported by Machado et al., who showed that pancreatic SPNs detected after 2000 were smaller than those diagnosed before [28]. However, the smaller tumor size in the MI group may have been a factor in deciding which surgical approach to follow, as previous studies have reported [16,29].

Our data also revealed a trend of an increased ratio of MI surgery (0 MI procedures in 10 cases before 2010; 4 MI surgeries out of 14 [29%] after 2010), which may be due to the progress of MI surgery skills, MI instruments, and the robotic system, as published by Cawich et al. [30]. However, the SPN locations ratio was similar before and after 2010 (head/body/tail; before 3/3/4, after 5/4/5).

No residual tumor was found after the operation, and all pathological reports indicated margin-free statuses in all cases in our study. This result suggests that MI and conventional operations deliver comparable long-term oncological results, as reported by Tan et al. [29]. Although some of the literature has discussed circumferential resection margins (CRM) in pathology reports of pancreatic cancer, there was no consensus on CRM for SPNs in our hospital. Patient enrollment commenced in 1998, and the perspective on CRM was introduced in 2006. Moreover, CRM was not addressed in earlier studies of SPNs due to the generally positive oncological outcomes associated with this condition [31,32].

The most commonly detected immunohistochemistry (IHC) stains in our study included CD10 (14 cases, 100% in CD10 stain cases), CD56 (7 cases, 100% in CD56 stain cases), and B-catenin (12 cases, 100% in B-catenin stain cases). As reported by Watanabe et al., vimentin, CD10, and CD56, are characteristically positive, and B-catenin was involved in the pathogenesis of SPN [6]. However, the histopathological features demonstrated no correlations in locations and prognoses in our study, like other reports [33].

In one case (4.2%), liver metastasis was found 41 months postoperatively, which matched the 2–10% postoperative recurrence after SPN radical resection reported previously [34]. Previous research has also indicated that the liver was the most common site of distant metastasis [21,35]. Despite the chemotherapy and transarterial chemoembolization (TACE) treatments that the patient received after liver metastasis, carcinomatosis still developed during the follow-up period, consistent with many reports showing a limited response to different regimens of chemotherapy [4,13,21]. Nevertheless, long-term survival was still observed; the patient survived for 159 months after surgery. This highlights the good oncological prognosis of SPN after surgery, which has been reported to range from 7 to 10 years after undergoing complete resection, even in patients with residual and disseminated diseases [10,13,21]. Sumida et al. reported that living donor liver transplantation for SPN with multiple liver metastatic lesions after complete resection was a possible therapeutic procedure with a >2 years of disease-free survival [35].

A 9-year-old female patient in our study suffered from jaundice after PD, and NASH was diagnosed through liver biopsy 18 months after PD. Though pediatric pancreatic tumors and PD in pediatric patients are rare, Sawai et al. reported a 10-year-old girl suffering from nonalcoholic fatty liver disease (NAFLD) developed after PD for SPN [36]. No previous reports describe NAFLD in children after PD due to its rarity. The mechanism of NAFLD or NASH remains uncertain, with a few reports hypothesizing that decreased exocrine function, deficiency in zinc, or bacterial translocation due to intestinal mucosal atrophy after PD play a role [36]. McGhee-Jez et al. reported an increased risk of post-PD NAFLD in women and adult patients with pancreatic cancer, shorter postoperative hospital stays, or higher preoperative BMIs [37]. Though the role of post-PD pancreatic enzyme supplementation is under debate, closely following patient status and pancreatic enzyme supplementation are suggested in pediatric patients after PD according to our experience [36,37]. We should also place more emphasis on psychological care in young patients who undergo pancreatic surgery.

Our data included a case of post-pancreatectomy diabetes in a patient who was at risk due to an extremely high BMI (43.75), as reported by Kwon JH. [38]. In our study, endocrine function was well preserved, even under aggressive resection (post-pancreatectomy diabetes, 4.2%), similar to other reports [39,40]. Some research reports 10–20% exocrine insufficiency after pancreatectomy, with steatorrhea or weight loss resolving after pancreatic enzyme supplementation [39,40].

According to the parenchyma preserving concept, we performed robotic (Xi)-assisted CP and gastropancreaticostomy with spleen preservation in a 26-year-old woman with a 3.2 cm tumor over the pancreatic body; no specific postoperative complications developed. No recurrence nor distant metastasis were found during the 17-month follow-up period. In our study, we observed no steatorrhea nor weight loss after pancreatectomy during the follow-up period. Despite some reports suggesting a parenchyma-preserving approach for SPN, due to its low malignant risk and possible decline in pancreatic endocrine and exocrine functions, oncologic outcomes were unclear due to the incremental risk of margin involvement. In our experience, parenchyma-preserving operations may be safe and feasible for treating pancreatic SPNs, but further research is required to clarify the benefits and drawbacks of parenchyma-preserving operations [21,27,39,40].

Our study has several limitations. Firstly, a notable challenge lies in the scarcity of high-quality, prospective, randomized controlled studies investigating the role of MI surgery in pancreatic SPN, largely due to its rare epidemiology. To address this gap, further studies are warranted, such as multi-center surveys or studies involving larger patient cohorts. These studies should aim to compare the efficacy of various surgical approaches through block or stratified randomization, thereby facilitating the acquisition of high-level evidence regarding the optimal management strategies for SPN.

Secondly, it is notable that our hospital lacks standardized preoperative evaluation protocols, encompassing diabetes profiles, imaging studies, and EUS reports. Particularly significant is the absence of a standardized immunohistochemistry stain in our hospital, considering the pathological variations identified in the research on SPN over the extended study period. Furthermore, our hospital lacks a standardized follow-up protocol for postoperative outcomes, including imaging, follow-up schedules, and diabetes profiles.

Thirdly, the extensive study period spanning more than 20 years poses challenges in evaluating not only the surgical techniques but also preoperative radiological diagnostics. This is particularly attributable to the advancements in MI techniques, MI equipment, and radiological facilities observed over this prolonged duration.

5. Conclusions

In conclusion, aggressive complete resection of pancreatic SPNs can be suggested based on our experience and according to the promising long-term survival achieved. Furthermore, this operation leads to no more complications than conventional surgery in treating SPNs of the pancreatic head and distal locations with good oncological outcomes. No histopathological correlations were found in prognoses and locations.

Author Contributions: Conceptualization, P.-Y.K. and W.-H.C.; data curation, P.-Y.K., S.-B.C., Y.-J.C., C.-Y.L., H.-T.L. and W.-H.C.; formal analysis, P.-Y.K.; investigation, P.-Y.K.; resources, S.-B.C., Y.-J.C., C.-Y.L., H.-T.L. and W.-H.C.; supervision, W.-H.C.; validation, P.-Y.K. and W.-H.C.; writing—original draft, P.-Y.K.; writing—review and editing, W.-H.C. All authors have read and agreed to the published version of the manuscript.

Funding: This research received no external funding.

Institutional Review Board Statement: The study was conducted in accordance with the Declaration of Helsinki and approved by the Institutional Review Board I & II of Taichung Veterans General Hospital (TCVGH-IRB No.: CE22251A) on 14 June 2022.

Informed Consent Statement: The informed consent requirement was waived due to the retrospective nature of the study.

Data Availability Statement: Data are unavailable due to privacy or ethical restrictions.

Acknowledgments: The authors are especially grateful to the colleagues, alliance, and biostatistics teams of Taichung Veterans General Hospital that kindly took part in the survey.

Conflicts of Interest: The authors declare no conflicts of interest.

References

1. Frantz, V.K. Atlas of Tumor Pathology. In *Tumors of the Pancreas*; US Armed Forces Institute of Pathology: Washington, DC, USA, 1959; pp. 32–33.
2. Hamoudi, A.B.; Misugi, K.; Grosfeld, J.L.; Reiner, C.B. Papillary epithelial neoplasm of pancreas in a child. Report of a case with electron microscopy. *Cancer* **1970**, *26*, 1126–1134. [CrossRef]
3. Hruban, R.H.; Pitman, M.B.; Klimstra, D.S. *Fascicle 6. Solid Pseudopapillary Neoplasms. Atlas of Tumor Pathology*; American Registry of Pathology and Armed Forces Institute of Pathology: Washington, DC, USA, 2007.
4. Reddy, S.; Cameron, J.L.; Scudiere, J.; Hruban, R.H.; Fishman, E.K.; Ahuja, N.; Pawlik, T.M.; Edil, B.H.; Schulick, R.D.; Wolfgang, C.L. Surgical management of solid-pseudopapillary neoplasms of the pancreas (Franz or Hamoudi tumors): A large single-institutional series. *J. Am. Coll. Surg.* **2009**, *208*, 950–959. [CrossRef]
5. Lima, C.A.; Silva, A.; Alves, C.; Alves, A., Jr.; Lima, S.; Cardoso, E.; Brito, E.; Macedo-Lima, M.; Lyra, D., Jr.; Lyra, P.; et al. Solid pseudopapillary tumor of the pancreas: Clinical features, diagnosis and treatment. *Rev. Da Assoc. Medica Bras. (1992)* **2017**, *63*, 219–223. [CrossRef]
6. Watanabe, Y.; Okamoto, K.; Okada, K.; Aikawa, M.; Koyama, I.; Yamaguchi, H. A case of aggressive solid pseudopapillary neoplasm: Comparison of clinical and pathologic features with non-aggressive cases. *Pathol. Int.* **2017**, *67*, 202–207. [CrossRef]
7. Papavramidis, T.; Papavramidis, S. Solid pseudopapillary tumors of the pancreas: Review of 718 patients reported in English literature. *J. Am. Coll. Surg.* **2005**, *200*, 965–972. [CrossRef]
8. Lüttges, J. Was ist neu? Die WHO-Klassifikation 2010 für Tumoren des Pankreas [What's new? The 2010 WHO classification for tumours of the pancreas]. *Pathologe* **2011**, *32*, 332–336. [CrossRef]
9. Law, J.K.; Ahmed, A.; Singh, V.K.; Akshintala, V.S.; Olson, M.T.; Raman, S.P.; Ali, S.Z.; Fishman, E.K.; Kamel, I.; Canto, M.I.; et al. A systematic review of solid-pseudopapillary neoplasms: Are these rare lesions? *Pancreas* **2014**, *43*, 331–337. [CrossRef] [PubMed]
10. Guo, N.; Zhou, Q.B.; Chen, R.F.; Zou, S.Q.; Li, Z.H.; Lin, Q.; Wang, J.; Chen, J.S. Diagnosis and surgical treatment of solid pseudopapillary neoplasm of the pancreas: Analysis of 24 cases. Canadian journal of surgery. *Can. J. Surg.* **2011**, *54*, 368–374. [CrossRef]
11. Virgilio, E.; Mercantini, P.; Ferri, M.; Cunsolo, G.; Tarantino, G.; Cavallini, M.; Ziparo, V. Is EUS-FNA of solid-pseudopapillary neoplasms of the pancreas as a preoperative procedure really necessary and free of acceptable risks? *Pancreatology* **2014**, *14*, 536–538. [CrossRef] [PubMed]
12. Sanfey, H.; Mendelsohn, G.; Cameron, J.L. Solid and papillary neoplasm of the pancreas. A potentially curable surgical lesion. *Ann Surg.* **1983**, *197*, 272–275. [CrossRef] [PubMed]
13. Martin, R.C.; Klimstra, D.S.; Brennan, M.F.; Conlon, K.C. Solid-pseudopapillary tumor of the pancreas: A surgical enigma? *Ann. Surg. Oncol.* **2002**, *9*, 35–40. [CrossRef]
14. Kang, C.M.; Kim, K.S.; Choi, J.S.; Kim, H.; Lee, W.J.; Kim, B.R. Solid pseudopapillary tumor of the pancreas suggesting malignant potential. *Pancreas* **2006**, *32*, 276–280. [CrossRef] [PubMed]
15. Hassan, I.; Celik, I.; Nies, C.; Zielke, A.; Gerdes, B.; Moll, R.; Ramaswamy, A.; Wagner, H.J.; Bartsch, D.K. Successful treatment of solid-pseudopapillary tumor of the pancreas with multiple liver metastases. *Pancreatology* **2005**, *5*, 289–294. [CrossRef]
16. Kang, C.M.; Choi, S.H.; Hwang, H.K.; Lee, W.J.; Chi, H.S. Minimally invasive (laparoscopic and robot-assisted) approach for solid pseudopapillary tumor of the distal pancreas: A single-center experience. *J. Hepatobiliary Pancreat. Sci.* **2011**, *18*, 87–93. [CrossRef] [PubMed]
17. De Sol, A.; Cirocchi, R.; Di Patrizi, M.S.; Boccolini, A.; Barillaro, I.; Cacurri, A.; Grassi, V.; Corsi, A.; Renzi, C.; Giuliani, D.; et al. The measurement of amylase in drain fluid for the detection of pancreatic fistula after gastric cancer surgery: An interim analysis. *World J. Surg. Oncol.* **2015**, *13*, 65. [CrossRef]
18. Sato, G.; Ishizaki, Y.; Yoshimoto, J.; Sugo, H.; Imamura, H.; Kawasaki, S. Factors influencing clinically significant delayed gastric emptying after subtotal stomach-preserving pancre-atoduodenectomy. *World J. Surg.* **2014**, *38*, 968–975. [CrossRef]
19. American Diabetes Association Diagnosis and classification of diabetes mellitus. *Diabetes Care* **2014**, *37*, S81–S90. [CrossRef] [PubMed]

20. Vassos, N.; Agaimy, A.; Klein, P.; Hohenberger, W.; Croner, R.S. Solid-pseudopapillary neoplasm (SPN) of the pancreas: Case series and literature review on an enigmatic entity. *Int. J. Clin. Exp. Pathol.* **2013**, *6*, 1051–1059.
21. Liu, M.; Liu, J.; Hu, Q.; Xu, W.; Liu, W.; Zhang, Z.; Sun, Q.; Qin, Y.; Yu, X.; Ji, S.; et al. Management of solid pseudopapillary neoplasms of pancreas: A single center experience of 243 consecutive patients. *Pancreatology* **2019**, *19*, 681–685. [CrossRef]
22. Mehrabi, A.; Hafezi, M.; Arvin, J.; Esmaeilzadeh, M.; Garoussi, C.; Emami, G.; Kössler-Ebs, J.; Müller-Stich, B.P.; Büchler, M.W.; Hackert, T.; et al. A systematic review and meta-analysis of laparoscopic versus open distal pancreatectomy for benign and malignant lesions of the pancreas: It's time to randomize. *Surgery* **2015**, *157*, 45–55. [CrossRef]
23. Cantisani, V.; Mortele, K.J.; Levy, A.; Glickman, J.N.; Ricci, P.; Passariello, R.; Ros, P.R.; Silverman, S.G. MR imaging features of solid pseudopapillary tumor of the pancreas in adult and pediatric patients. AJR Am. *J. Roentgenol.* **2003**, *181*, 395–401. [CrossRef] [PubMed]
24. Wang, D.B.; Wang, Q.B.; Chai, W.M.; Chen, K.M.; Deng, X.X. Imaging features of solid pseudopapillary tumor of the pancreas on multi-detector row computed tomography. World J. *Gastroenterol.* **2009**, *15*, 829–835.
25. Kartal, E.; Schmidt, T.S.B.; Molina-Montes, E.; Rodríguez-Perales, S.; Wirbel, J.; Maistrenko, O.M.; Akanni, W.; Alhamwe, B.A.; Alves, R.J.; Carrato, A.; et al. A faecal microbiota signature with high specificity for pancreatic cancer. *Gut* **2022**, *71*, 1359–1372. [CrossRef] [PubMed]
26. Boicean, A.; Ichim, C.; Todor, S.B.; Anderco, P.; Popa, M.L. The Importance of Microbiota and Fecal Microbiota Transplantation in Pancreatic Disorders. *Diagnostics* **2024**, *14*, 861. [CrossRef] [PubMed]
27. Uguz, A.; Unalp, O.V.; Akpinar, G.; Karaca, C.A.; Oruc, N.; Nart, D.; Yilmaz, F.; Aydin, A.; Coker, A. Solid pseudopapillary neoplasms of the pancreas: Case series with a review of the literature. *Turk. J. Gastroenterol.* **2020**, *31*, 930–935. [CrossRef]
28. Machado, M.A.C.; Bacchella, T.; Jukemura, J.; Almeida, J.L.; Cunha, J.E. Solid pseudopapillary neoplasm of the pancreas: Distinct patterns of onset, diagnosis, and prognosis for male versus female patients. *Surgery* **2008**, *143*, 29–34. [CrossRef] [PubMed]
29. Tan, H.L.; Syn, N.; Goh BK, P. Systematic Review and Meta-analysis of Minimally Invasive Pancreatectomies for Solid Pseudopapillary Neoplasms of the Pancreas. *Pancreas* **2019**, *48*, 1334–1342. [CrossRef] [PubMed]
30. Cawich, S.O.; Kluger, M.D.; Francis, W.; Deshpande, R.R.; Mohammed, F.; Bonadie, K.O.; Thomas, D.A.; Pearce, N.W.; Schrope, B.A. Review of minimally invasive pancreas surgery and opinion on its incorporation into low volume and resource poor centres. *World J. Gastrointest. Surg.* **2021**, *13*, 1122–1135. [CrossRef] [PubMed]
31. Demir, I.E.; Jaeger, C.; Schlitter, A.M.; Konukiewitz, B.; Stecher, L.; Schorn, S.; Tieftrunk, E.; Scheufele, F.; Calavrezos, L.; Schirren, R.; et al. R0 Versus R1 Resection Matters after Pancreaticoduodenectomy, and Less after Distal or Total Pancreatectomy for Pancreatic Cancer. *Ann. Surg.* **2018**, *268*, 1058–1068. [CrossRef]
32. Safi, S.-A.; Alexander, A.; Neuhuber, W.; Haeberle, L.; Rehders, A.; Luedde, T.; Esposito, I.; Fluegen, G.; Knoefel, W.T. Defining distal splenopancreatectomy by the mesopancreas. *Langenbecks Arch. Surg.* **2024**, *409*, 127. [CrossRef]
33. Tasar, P.; Kilicturgay, S.A. Solid pseudopapillary neoplasms of the pancreas: Is there a factor determining the prognosis? Experience of a single institution. *Medicine* **2022**, *101*, e30101. [CrossRef] [PubMed]
34. Huffman, B.M.; Westin, G.; Alsidawi, S.; Alberts, S.R.; Nagorney, D.M.; Halfdanarson, T.R.; Mahipal, A. Survival and Prognostic Factors in Patients With Solid Pseudopapillary Neoplasms of the Pancreas. *Pancreas* **2018**, *47*, 1003–1007. [CrossRef] [PubMed]
35. Sumida, W.; Kaneko, K.; Tainaka, T.; Ono, Y.; Kiuchi, T.; Ando, H. Liver transplantation for multiple liver metastases from solid pseudopapillary tumor of the pancreas. *J. Pediatr. Surg.* **2007**, *42*, e27–e31. [CrossRef] [PubMed]
36. Sawai, T.; Zuo, S.; Terai, T.; Nishiwada, S.; Nakagawa, K.; Nagai, M.; Akahori, T.; Kanehiro, H.; Sho, M. Nonalcoholic fatty liver disease developed after pancreatoduodenectomy for solid pseudopapillary neoplasm in a 10-year-old girl: A case report. *Surg. Case Rep.* **2022**, *8*, 61. [CrossRef] [PubMed]
37. McGhee-Jez, A.E.; Chervoneva, I.; Yi, M.; Ahuja, A.; Nahar, R.; Shah, S.; Loh, R.; Houtmann, S.; Shah, R.; Yeo, C.J.; et al. Nonalcoholic Fatty Liver Disease After Pancreaticoduodenectomy for a Cancer Diagnosis. *J. Pancreat. Cancer* **2021**, *7*, 23–30. [CrossRef] [PubMed]
38. Kwon, J.H.; Kim, S.C.; Shim, I.K.; Song, K.B.; Lee, J.H.; Hwang, D.W.; Park, K.-M.; Lee, Y.-J. Factors Affecting the Development of Diabetes Mellitus After Pancreatic Resection. *Pancreas* **2015**, *44*, 1296–1303. [CrossRef] [PubMed]
39. Shyr, B.S.; Wang, S.E.; Chen, S.C.; Shyr, Y.M.; Shyr, B.U. Pancreatic head sparing surgery for solid pseudopapillary tumor in patients with agenesis of the dorsal pancreas. *J. Chin. Med. Assoc.* **2022**, *85*, 981–986. [CrossRef]
40. Li, Y.-Q.; Pan, S.-B.; Yan, S.-S.; Jin, Z.-D.; Huang, H.-J.; Sun, L.-Q. Impact of parenchyma-preserving surgical methods on treating patients with solid pseudopapillary neoplasms: A retrospective study with a large sample size. *World J. Gastrointest. Surg.* **2022**, *14*, 174–184. [CrossRef]

MDPI AG
Grosspeteranlage 5
4052 Basel
Switzerland
Tel.: +41 61 683 77 34

Medicina Editorial Office
E-mail: medicina@mdpi.com
www.mdpi.com/journal/medicina

www.ingramcontent.com/pod-product-compliance
Lightning Source LLC
LaVergne TN
LVHW071941160726
843515LV00011B/2668